T0243044

Prodrug Design

Prodrug Design

Perspectives, Approaches and Applications in Medicinal Chemistry

Vivekkumar K. Redasani

Sanjay B. Bari

AMSTERDAM • BOSTON • HEIDELBERG • LONDON
NEW YORK • OXFORD • PARIS • SAN DIEGO
SAN FRANCISCO • SINGAPORE • SYDNEY • TOKYO
Academic Press is an imprint of Elsevier

Academic Press is an imprint of Elsevier
125, London Wall, EC2Y 5AS
525 B Street, Suite 1800, San Diego, CA 92101-4495, USA
225 Wyman Street, Waltham, MA 02451, USA
The Boulevard, Langford Lane, Kidlington, Oxford OX5 1GB, UK

Notices
Knowledge and best practice in this field are constantly changing. As new research and experience broaden our understanding, changes in research methods, professional practices, or medical treatment may become necessary.

Practitioners and researchers must always rely on their own experience and knowledge in evaluating and using any information, methods, compounds, or experiments described herein. In using such information or methods they should be mindful of their own safety and the safety of others, including parties for whom they have a professional responsibility.

To the fullest extent of the law, neither the Publisher nor the authors, contributors, or editors, assume any liability for any injury and/or damage to persons or property as a matter of products liability, negligence or otherwise, or from any use or operation of any methods, products, instructions, or ideas contained in the material herein.

ISBN: 978-0-12-803519-1

British Library Cataloguing-in-Publication Data
A catalogue record for this book is available from the British Library.

Library of Congress Cataloging-in-Publication Data
A catalog record for this book is available from the Library of Congress.

For Information on all Academic Press publications
visit our website at http://store.elsevier.com/

This book has been manufactured using Print On Demand technology.

DEDICATION

TO ALL SCIENTIST WORKING IN MEDICINAL CHEMISTRY

CONTENTS

PREFACE

The role of medicinal chemistry in the discovery of drugs and later medicines has now become an integral part of science. Discovery in any area is a never-ending process. However, of the several fields of working, there is a need to keep updating the developments in medicinal chemistry. This is an approach where, apart from therapeutic effects, the unwanted effects are also in need of study. This book is a means of communication between students and young scientists working in the area of prodrugs. It will provide a framework for learning the necessary skills and applying them powerfully.

A growing demand in medicinal chemistry encourages us to proceed in publishing this book. Apart from the researchers, great care has been taken to keep in view the needs of students and students from various universities across the globe will benefit from reading this book. The book contains a great deal of valuable information that is supported by illustrations. A proper classification system dependent upon the therapeutic class has ensured that all aspects of medicinal chemistry are covered in relation to prodrugs. This gives an idea of how widely the concept of prodrugs can be applied. The simplified methods of preparation of derivatives should create an interest in budding researchers that will motivate them towards this field whilst also removingany fears students may have of organic and medicinal chemistry. Current and updated examples ensure that the contents are fully up-to-date.

The book contents are divided into several partsencompassing an introduction, drug discovery and the reported work on prodrugs. Following the Introduction, the book covers the concept, types, classes, and approaches in a logical order. A chapter on application relates the present state of affairs and highlights the current need for prodrugs. Following on from its role in drug discovery, an updated reported work is discussed, offering a new sight into prodrug research. All chapters have also been thoroughly referenced.

Given the approach that we have taken, we are confident this book will appeal to academic and industrial researchers in chemistry medicine and pharmacy, as well as teachers of the subjects at this level.

We are grateful for and acknowledge our management and principal for their motivational support and spirit of inspiration. We are also thankful to all those individuals who contributed directly or indirectly in the completion of this work.

Dr. Vivekkumar K Redasani
Dr. Sanjay B Bari

CHAPTER 1

Introduction

1.1 BACKGROUND

The goal of drug design is to correlate biological activity and physicochemical properties. However, drug design should not merely be aimed at increased pharmacological activity. Instead, it is more desirable to achieve a better ratio between the activity and toxicity of drugs. The majority of drugs in use today have been developed by a traditional empirical approach where several thousands of substances have been screened for biological activity and of these only one may eventually become a new drug (Hansch et al., 2005).

1.2 DRUG DEVELOPMENT

The process of taking a new drug from concept to clinic and eventually to commercialization involves several steps that traditionally occur in series. The discovery and development of a new drug start from the identification and validation of the target protein or receptor. The drug development practice from the target identification to the final product is a time- and money-consuming process (DiMasi et al., 2003). The average cost of taking a compound from the discovery stage to commercialization is estimated to exceed US$800 million, with an average development timespan of ~15 years. Finally, perhaps only one of 10 clinically studied drug candidates may reach the market and, as a whole, the probability for each synthesized compound reaching the market is thus *only one* in a million. The reasons for this poor percentage of success in the drug development process are poor drug properties, such as solubility and permeability which are amongst the main causes for failure, along with toxicity as an additional factor (Borchardt et al., 2006).

Given these frightening statistics, pharmaceutical companies are under constant pressure to streamline the drug development process. The recent craze of drug patent expirations and subsequent generic competition for blockbuster drugs has further aggravated the need for

Prodrug Design. DOI: http://dx.doi.org/10.1016/B978-0-12-803519-1.00001-5

efficient strategies to screen, identify, and optimize lead compounds for development.

High-throughput screening (HTS) and combinatorial chemistry methodologies have been developed in the past two decades, to synthesize a vast number of compounds using limited resources. Large numbers of molecules are emerging as a result of these methodologies. Several complementary *in silico* and *in vitro* strategies have also emerged to screen these compounds and assess their potential to become lead candidates. Compounds emerging as "hits" from these screening processes are characterized further and tested *in vivo* for safety and efficacy. This phase of preclinical drug development requires that the compound be formulated into a dosage form that can be used to administer the drug. Although simple conventional formulation technologies are preferred at this stage, such technologies are often not applicable for challenging compounds, such as those exhibiting poor aqueous solubility (Chaubal, 2004). However, many such potential drug candidate leads exhibit high affinities towards a variety of molecular targets, such as receptors, enzymes, etc. Not all are going to become real drug candidates due to their inherent physicochemical properties. A drug can only exert a desired pharmacological effect if it reaches its site of action.

Even so, the modern discovery technologies, such as HTS and combinatorial chemistry, can produce novel lead structures with high pharmacological potency, but the physicochemical and biopharmaceutical aspects of the initial leads have frequently been neglected. This can lead to drug candidates with poor drug-like properties that face significant problems later in drug development (Venkatesh and Lipper, 2000).

1.3 THE DRUG DISCOVERY PROCESS

Drug discovery has evolved from rational drug design, which was slow and time-consuming, to HTS and combinatorial chemistry technologies that can result in up to several compounds being synthesized per chemist per year. Drug discovery productivity has been accelerated further by the availability of information from the genomic database, which provides new biological targets for drug development. Under such circumstances the appropriate screening of a vast number of compounds becomes crucial for the success of the drug discovery program.

Over the past four decades, the pharmaceutical industry has experienced a sharp swing towards discovery and development of new drug molecules. A significant variation is seen in order to improve or eliminate drawbacks related to physicochemical properties which determine the pharmacokinetic behavior, and pharmaceutical and biological performance of already existing drugs.

1.3.1 Goal of Drug Discovery

The goal of many researchers is the design of a drug that hits only the drug target, while minimizing drug exposure to other sites in the body, thus minimizing toxicity. This targeting idea has been the goal of numerous researchers via the use of prodrugs. The frustrations that drug discovery process teams encounter with the new bigger, more complex drug molecule candidates have renewed interest in this novel problem-solving technique and have led to some significant recent commercial successes. To overcome the undesirable physicochemical, biological, and organoleptic properties of some marketed drugs, the development of new chemical compounds with satisfactory efficacy and safety has to be made. However, this is a very expensive and time-consuming process.

1.3.2 Constraints in Drug Discovery

When a new chemical entity has some barrier/limitation to utility, it may not be developed as a therapeutic agent. For example, the drug may be water-soluble, making it difficult for a safe injectable dosage form to be developed for human use. Another limitation might be that the drug, although effective if given by injection, cannot be absorbed through the gastrointestinal tract. This may be because it is very polar to cross the cell linings the gastrointestinal tract or because the chemical is metabolized by enzymes present in these cells or in liver, thus preventing the drug from reaching the systemic circulation.

1.4 CURRENT SCENARIO IN PRODRUG RESEARCH

At present, revolutionary steps in the drug discovery and development process have been recognized due to the dawn of pharmacogenomics, proteomics, bioinformatics, HTS, virtual screening, *de novo* design, *in silico* ADME screening, and structure-based drug design. Various computational methods are utilized for the design of high-affinity receptor or enzyme binders, either through virtual computer screening of compound libraries or through design and synthesis of novel entities. These computational methods are very well known in evaluating the target structures for possible binding to active sites, generating candidate chemical structures, assigning their drug-likeness properties, docking chemical structures to active sites and optimizing the candidate molecules to improve their binding properties. However, these computational methods, along with the techniques used for developing a drug candidate, can provide good *in vitro* drug activity but cannot be extrapolated to good *in vivo* drug activity unless a drug candidate has good bioavailability and a desirable duration of action. Therefore, a growing awareness of finding alternative approaches as determinants of *in vivo* drug therapeutic activity has led the drug industry to pursue the prodrug approach as a prime priority.

The rationale of using prodrugs is to achieve optimum pharmacokinetic properties and to enhance greatly the selectivity of a drug for its target's active site. Utilizing the prodrug approach, the potential drug candidate is made available with improved properties. A harmony has been reached that the prodrug approach is a promising and well-established strategy to develop new entities with superior efficacy and selectivity and reduced toxicity. Therefore, an optimized therapeutic outcome can be accomplished. Approximately a tenth of all worldwide marketed medicines can be classified as prodrugs, and in 2008 alone, a third of all approved drugs with low molecular weights were prodrugs. This fact, without doubt, indicates the great success of the prodrug approach.

Traditionally, the prodrug approach was aimed at improving the physicochemical properties of drugs via covalently attaching the drug moiety to a nontoxic chemical species, the *promoiety*. The prodrug is intended to interconvert within the body by specific enzymes to liberate the parent active drug. The prodrug can be hydrophilic, aiming to increase solubility in the gastrointestinal tract or aiming to enhance membrane permeability. Such prodrugs suffer from nonspecific activation at sites other than the active site, resulting in related toxicities and low bioavailability. The molecular revolution and the advance in computational chemistry in recent years, and the ample increase in knowledge of the structures and functions of enzymes and transporters have created a new era of prodrugs which are termed *targeted prodrugs*. Researchers have now shifted from synthesizing classical prodrugs to designing prodrugs for specific targeting of enzymes and transporters, thus increasing bioavailability and reducing toxicity, and therefore achieving a better therapeutic profile of drug candidate (Karaman, 2014).

1.5 NEED OF THE STUDY

According to Professor Takeru Higuchi, "Drugs need to be design with delivery in mind" (quoted in the mid-1970s). This paradigm is very important in terms of drug discovery, and it should be borne in mind at all stages of drug development. Drug metabolism pharmacokinetic research has helped to develop drugs with suitable properties. Novel formulation approaches also help overcome basic delivery limitations caused by problematic physicochemical or biochemical properties. Furthermore, the prodrug approach is an integral approach to drug discovery (Teruko, 2011).

Thus, instead of synthesizing new compounds, which is a time-consuming and too costly affair, the designing of derivatives of existing clinically used drugs is definitely an interesting and promising area of research. Moreover, as the metabolic profile of the liberated parent drug (after cleavage of the derivative in the body) would be already known, it could be advantageous to design prodrug derivatives of parent compounds. Prodrug design comprises an area of research devoted to optimize drug delivery, where the pharmacologically inactive species required transformation within the body in order to release the parent active drug (Halen et al., 2009).

In the past, the prodrug approach was viewed as a last resort after all other ways were exploited, whereas, now-a-days the prodrug approach is being considered in the very early stages of the drug discovery and development process. While the traditional prodrug approach was focused on altering various physiochemical parameters, the modern computational approach considers using a design of targeted prodrugs to certain enzymes or transporters or being interconverted to their parent drugs without a metabolic activation process. With the possibility of designing prodrugs with different linkers, the rate of release of the parent drug will be controlled and the drug's moiety, responsible for a bitter sensation, will be blocked (Karaman, 2014).

This growing demand in the drug discovery process encouraged us to produce this particular edition of this book. Great care has been taken to fulfill the needs of students and researchers alike. From the student point of view, the contents will provide precise data and cover mainly the prescribed syllabus of various universities. For the different courses, including basic sciences and pharmaceutical sciences, where there is a need to study the drug discovery process, this book might be a beneficial one. From the research point of view, the examples available enforce one to find a new sight. The realization of the need for and importance of this book promoted us to upgrade the previous edition.

REFERENCES

Borchardt, R.T., Kerns, E.H., Hageman, M.J., Thakker, D.R., Stevens, J.L., 2006. Optimizing the "Drug Like" Properties of Leads in Drug Discovery. Springer Science and Business Media, New York, NY.

Chaubal, M.V., 2004. Application of formulation technologies in lead candidate selection and optimization. Drug Discov. Today 9 (14), 603–609.

DiMasi, J.A., Hansen, R.W., Grabowski, H.G., 2003. The price of innovation: new estimates of drug development costs. J. Health Econ. 22, 15185.

Halen, P.K., Murumkar, P.R., Giridhar, R., Yadav, M.R., 2009. Prodrug designing of NSAIDs. Mini Rev. Med. Chem 9 (1), 124–139.

Hansch, C., Sammes, P.G., Taylor, J.B., 2005. Comprehensive Medicinal Chemistry: "Principles of Pharmacokinetics and Metabolism", vol. 5. Pergamon Press, Oxford England, pp. 122–133.

Karaman, R., 2014. Prodrugs-current and future drug development strategy. Int. J. Med. Pharm. Case Reports 1 (2), 58–63.

Teruko, I., 2011. Prodrug approach in current drug discovery. Drug Metab. Pharmacokinet 26 (4), 307–308.

Venkatesh, S., Lipper, R.A., 2000. Role of the development scientist in compound lead selection and optimization. J. Pharm. Sci. 89, 145–154.

CHAPTER 2

Concept of Prodrug

2.1 CONCEPT OF PRODRUG

Many therapeutic drugs have undesirable properties that may become pharmacological, pharmaceutical, or pharmacokinetic barriers in clinical drug application. Among the various approaches to minimize the undesirable drug properties, while retaining the desirable therapeutic activity, the chemical approach using drug derivatization offers perhaps the highest flexibility and has been demonstrated as an important means of improving drug efficacy (Han, 2000). The prodrug approach, a chemical approach using reversible derivatives, can be useful in the optimization of the clinical application of a drug. The prodrug approach gained attention as a technique for improving drug therapy in the early 1970s. Numerous prodrugs have been designed and developed since then to overcome pharmaceutical and pharmacokinetic barriers in clinical drug application, such as low oral drug absorption, lack of site specificity, chemical instability, toxicity, and poor patient acceptance (bad taste, odor, pain at injection site, etc.) (Stella, 1975). It is justified by the fact that once the barrier to the use of the parent compound has been overcome, these temporary forms can be converted to the free parent compound that can exert its pharmacological activity. The prodrug approach can be illustrated as shown in Figure 2.1.

2.2 UNDESIRABLE PROPERTIES ASSOCIATED WITH DRUG MOLECULES

There are many undesirable properties associated with drug molecules. These properties demonstrate restrictions in its use.

Physical properties include:

- Poor aqueous solubility
- Low lipophilicity
- Chemical instability.

Prodrug Design. DOI: http://dx.doi.org/10.1016/B978-0-12-803519-1.00002-7

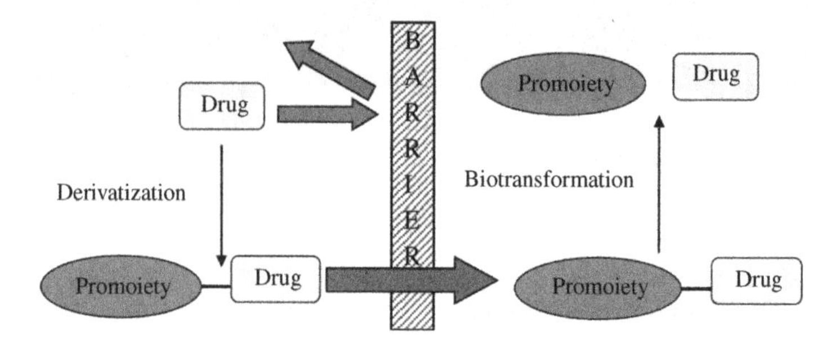

Figure 2.1 Schematic illustration of prodrug concept.

Pharmacokinetic properties include:

- Poor distribution across biological membranes
- Good substrate for first-pass metabolism
- Rapid absorption/excretion when a long-term effect is desired.

The poor drug delivery problem can be overcome by one of the following approaches:

1. By designing appropriate dosage form.
2. Preparing a new drug analog of the original drug.
3. Exploitation of bioreversible chemical derivatization, that is, the*prodrug approach.*

Several barriers of nonpharmacokinetic and pharmacodynamic origin may prevent a drug from reaching the desired target. This includes pathological limitations, such as toxicity, high incidence of side effects, pharmaceutical limitations, psychological limitations, and also economic barriers. Most of these limitations can be overcome by the prodrug approach (Higuchi and Davis, 1970).

During the last two decades, there has been a steady improvement in the physicochemical, biopharmaceutical, and/or pharmacokinetic properties of pharmacologically active compounds by the implementation of a prodrug strategy.

The basic aim of prodrug design is to mask undesirable drug properties, such as low solubility in water or lipid membranes, low target selectivity, chemical instability, undesirable taste, irritation or pain after local administration, presystemic metabolism, and toxicity.

2.3 PRODRUG DESIGN: PAST TO PRESENT

The first intentionally designed derivative of a drug molecule was introduced as long ago as 1899 by Schering, this was methenamine (or hexamine). Methenamine releases antibacterial formaldehyde along with ammonium ions in acidic urine and serves as a good example of a site-selective prodrug. At the same time, Bayer introduced aspirin (acetylsalicylic acid) as a less irritating form of the anti-inflammatory agent sodium salicylate. However, it remains unclear whether aspirin is a true prodrug or not. Although it was intended to work as a prodrug, aspirin inhibits irreversibly cyclooxygenase, the enzyme responsible for the formation of the key biological mediators, prostaglandins and thromboxanes. The parent drug, salicylic acid, is a weak reversible inhibitor of cyclooxygenase and thus cannot be considered a prodrug. However, aspirin is rapidly hydrolyzed in the intestinal wall and liver, as well as in the blood, to salicylic acid, which means that it acts, in fact, like a prodrug.

Later, Bayer introduced another prodrug, the antibiotic prontosil, in 1935. However, prontosil was not intentionally developed as a prodrug, because very soon it was found to release an active agent, *para*-aminophenylsulfonamide, by reductive enzymes. This led to the discovery of the widely used antibacterials, sulfonamides. Similarly, Roche discovered the prodrug activity of the antitubercular agent isoniazid more than 40 years after its introduction in 1952. Today, it is known that the bioactivation of isoniazid is catalyzed by the peroxidase enzyme and the generated reactive inhibits the biosynthesis of mycolic acid required for the mycobacterial cell wall.

Sometimes, unintentionally developed prodrugs can reveal a less appealing truth of the drug under the development. Heroin (diacetylmorphine), was marketed during the years 1898–1910 as a nonaddictive morphine substitute to suppress cough and cure morphine addictions. However, later it became known that, in fact, heroin is rapidly metabolized into morphine after oral administration.

There has been an explosive boost since the 1960s in the use of prodrugs in drug discovery and development. From the start of the twenty-first century, when property-based drug design became an essential part of drug discovery and development, it has been a time of real breakthroughs in prodrugs. Thus, day by day the number of prodrugs undergoing clinical trials is increasing and prodrug research is now becoming an integral part of drug design.

One interesting example of these blockbuster prodrugs is lisdexamfetamine dimesylate, the L-lysine prodrug of the psychomotor stimulant dextroamphetamine. It was designed to have less abuse potential than other amphetamines due to the slower release of the active parent drug if inhaled or injected. Another best-selling group of prodrugs is the proton pump inhibitors omeprazole and its analogs, which are site-selectively bioactivated to their active species in the acidic parietal cells of the stomach.

2.4 DEFINITIONS OF PRODRUG

The term "prodrug" or "proagent" was first introduced by Albert (1958) to signify pharmacologically inactive chemical derivatives that could be used to alter the physicochemical properties of drugs, in a temporary manner, to increase their usefulness and/or to decrease associated toxicity.

The concept given by Albert was later extended by *Harper*, who coined the term *drug latentiation*, which implies a time lag element or component. Later, the concept of prodrug and latentiated drugs for solving various problems was attempted and the definition of drug latentiation was extended to include nonenzymatic regeneration of parent compounds. Thus, such compounds have also been called "latentiated drugs," "bioreversible derivatives," and "congeners," but "prodrug" is now the most commonly accepted term (Roche, 1977; Sinkula and Yalkowsky, 1975).

Usually, the use of the term implies a covalent link between a drug and a chemical moiety, though some authors also use it to characterize some forms of salts of the active drug molecule. Although there is no strict universal definition for a prodrug itself, and the definition may vary from author to author, generally prodrugs can be defined as pharmacologically inert chemical derivatives that can be converted *in vivo* to the active drug molecules, enzymatically or non-enzymatically, to exert a therapeutic effect. Ideally, the prodrug should be converted to the original drug as soon as the goal is achieved, followed by the subsequent rapid elimination of the released derivatizing group (Stella et al., 1985; Banerjee and Amidon, 1985).

In simplified terms, prodrugs are masked forms of active drugs that are designed to be activated after an enzymatic or chemical reaction once they have been administered into the body. Prodrugs are considered to be inactive or at least significantly less active than the released drugs; therefore, salts of active agents and drugs, whose metabolites contribute to the overall pharmacological response, are not included in these definitions.

2.5 RATIONALE FOR THE USE OF PRODRUGS

The rationale behind the use of prodrugs is generally to optimize the "drug-like" properties (ADME, absorption, distribution, metabolism, and excretion) because they can cause considerable problems in subsequent drug development, if unfavorable. The prodrug approach is a very versatile strategy to increase the utility of pharmacologically active compounds, because one can optimize any of the ADME and toxicity properties as well as prolong the commercial lifecycle of potential drug candidates. In addition, the prodrug strategy has been used to increase the selectivity of drugs for their intended target. This can not only improve the efficacy of the drug but also decrease systemic and/or unwanted tissue/organ-specific toxicity.

A drug can only exert a desired pharmacological effect if it reaches its site of action. The three major phases involved in the drug receptor interaction or biological bioavailability of drug include:

- the pharmaceutical phase
- the pharmacokinetic phase and
- the pharmacodynamic phase.

Many barriers which limit drugs' ability to reach a desired target organ and the subsequent receptor site are considered of pharmacokinetic origin as shown in Figure 2.2.

Besides these, barriers of nonpharmacokinetic and pharmacodynamic origin may also prevent a drug from reaching the desired target. It includes pathological limitations such as toxicity, high incidence of side effects and teratogenicity, pharmaceutical limitations such as chemical instability of product or formulation, psychological limitations such as unpleasant taste, pain at injection site and cosmetic damage to the patient, and economic barriers.

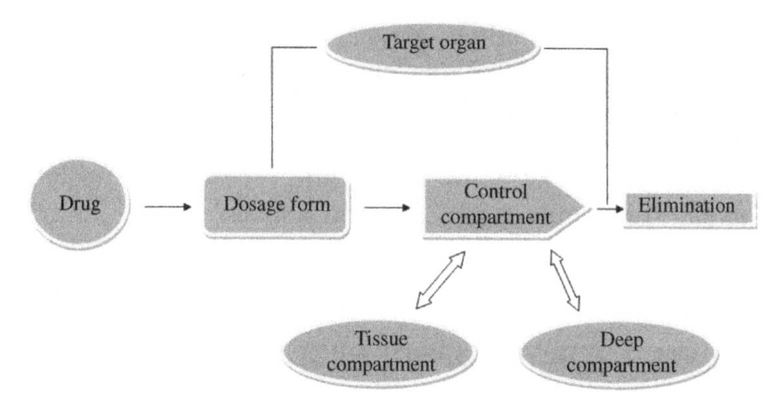

Figure 2.2 Pharmacokinetic model for a typical drug.

Most of these limitations can be overcome by the prodrug approach, but after overcoming the various barriers, the prodrug should rapidly convert into active moiety after reaching the target site. Awareness that the onset, intensity, and duration of drug action are greatly affected by the physicochemical properties of drugs has promoted the emergence of various theoretical and predictive models for drug design and evaluation. The design of an efficient, stable, safe, acceptable, and aesthetic way to target a drug to its site of action while overcoming various physical, chemical, and social barriers is certainly an area where the utilization of the prodrug approach holds great potential.

The *pharmaceutical phase* can be considered as the phase of development which involves the identification of a new chemical entity with measured or proposed therapeutic potential and its incorporation into a drug delivery system. The delivery system may be one of the traditional forms, such as tablets, capsules, injections and creams/ointments as well as the new drug delivery modes, such as liposomes, implants, etc.

The *pharmacokinetic phase* can be considered as the phase involving absorption, distribution, metabolism, and excretion of the drug. The pharmacokinetic studies provide valuable information regarding the *in vivo* properties of a drug's limitations, such as poor absorption, too rapid elimination, and presystemic metabolism. If these properties can be related back to the physicochemical and dosage form properties of the system, then corrections will require prodrug interventions.

2.6 TARGETED PRODRUG DESIGN

Classical prodrug design often represents a nonspecific chemical approach to mask undesirable drug properties, such as limited bioavailability, lack of site specificity, and chemical instability. On the other hand, targeted prodrug design represents a new strategy for directed and efficient drug delivery. Particularly, targeting the prodrugs to a specific enzyme or a specific membrane transporter, or both, has potential as a selective drug delivery system.

Prodrugs can be designed to target specific enzymes or carriers by considering enzyme–substrate specificity or carrier–substrate specificity in order to overcome various undesirable drug properties. This type of "targeted-prodrug" design requires considerable knowledge of particular enzymes or carrier systems, including their molecular and functional characteristics (Han, 2000).

2.7 DOUBLE PRODRUG CONCEPT

The prodrug approach is commonly practiced to improve drug delivery and drug targeting. the target-specific cleavage mechanism is followed in a prodrug design to encourage site-specific drug delivery. However, it will not serve the purpose if it is not able to reach the target. Also, stability problems are observed in prodrugs involving chemical release of active drug. These problems can be improved through a double prodrug approach in which the enzymatic release mechanism is essential prior to the spontaneous release of the parent compound.

2.8 STEPS IN PRODRUG DESIGN

- Identification of the drug delivery problem
- Identification of desired physicochemical properties
- Selection of transport moiety which will give the prodrug desired transport properties so that it can be readily cleaved in the desired biological compartment.

2.8.1 Hard Drugs

Drugs are sometimes divided into "hard drugs" and "soft drugs." Hard drugs are "nonmetabolizable drugs" or drugs which are metabolized to biologically active metabolites. The metabolites of hard drugs are frequently toxic oxidation products. Soft drugs are drugs which are characterized by a predictable and controllable *in vivo* destruction (i.e., metabolism) to nontoxic products after they have achieved their therapeutic role.

Similarly "hard compounds" can be defined as compounds which do not degrade in the environment or compounds which do it very slowly. Thus, these compounds will lead to progressive pollution of the environment. An example of a hard compound is the insecticide DDT.

2.8.2 Soft Drugs

"Soft compounds" can be defined as biologically active compounds which are readily degraded to nontoxic and biologically inactive degradation products in the environment. The purpose of this is to design, synthesize, and test soft drugs and soft environment-friendly compounds.

2.9 OBJECTIVES IN PRODRUG RESEARCH

There are three basic, overlapping objectives in prodrug research:

- Pharmaceutical
- Pharmacokinetic
- Pharmacodynamic.

2.9.1 Pharmaceutical Objectives

The pharmaceutical objectives are to improve solubility, chemical stability, and organoleptic properties; to decrease irritation and/or pain after local administration; and to reduce problems related to the pharmaceutical technology of the active agent.

Pharmaceutical scientists are often faced with serious formulation problems resulting from poor solubility, insufficient chemical stability, or poor organoleptic properties (bad smell or bitterness endangering patient compliance). While pharmaceutical technology can solve such problems in favorable cases (e.g., by improving the solubility of cyclosporine), success may be uncertain and time-consuming to achieve. Instead of waiting for an uncertain and delayed pharmaceutical solution to a problem of solubility or stability, it is better to take advantage of a prodrug strategy for an early solution. A representative example of a prodrug solution to a solubility problem is afforded by dapsone. It aimed at improving the water solubility of this drug by derivatizing into a number of prodrugs, creating an amido bridge between dapsone and an amino acid. Good water solubilities were indeed contributed by the more polar pro-moieties (i.e., glycyl, alanyl, and lysyl). Thus, solubility is one of the main factors influencing oral absorption. Increasing solubility is a pharmacokinetic as well as a pharmaceutical objective.

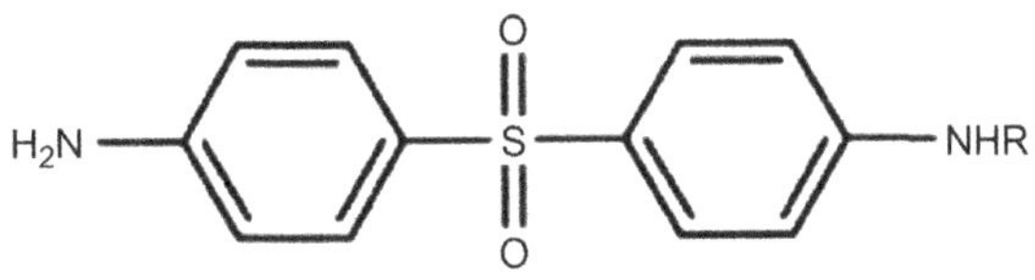

(Dapsone R = H; Prodrugs R = glycyl, alanyl, leucyl, lysyl, phenylalanyl)

2.9.2 Pharmacokinetic Objectives

Having a pharmacokinetic objective in the prodrug design serves as a very important approach. It facilitates in several ways: improvement in absorption (by oral and nonoral routes), decreasing presystemic metabolism to improve time profile, improvements due to change in formulation, and increase in organ/tissue-selective delivery of the active agent.

Foremost among these is a need to improve oral bioavailability, be it by improving the oral absorption of the drug and/or by decreasing its presystemic hepatic metabolism. Achieving improved oral absorption by a prodrug strategy is a frequent rationale in marketed prodrugs. Other objectives are to improve absorption by parenteral (nonenteral, e.g., dermal, ocular) routes and to increase the duration of action of the drug by having a slow metabolic release. This in turn finally achieves the organ/tissue-selective delivery of an active agent. Some of these objectives are exemplified with clinically successful prodrugs.

Zanamivir is a highly hydrophilic drug administered in aerosol form. The active agent is oseltamivir acid, which also shows very high *in vitro* inhibitory efficacy toward the enzyme but low oral bioavailability due to its high polarity. To circumvent this problem, it is marketed as its ethyl ester prodrug (oseltamivir), which undergoes rapid enzymatic hydrolysis and produces high and sustained plasma levels of the active agent.

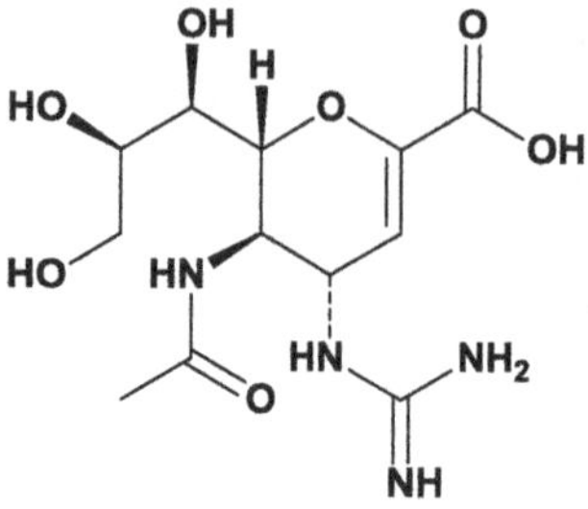

Zanamivir

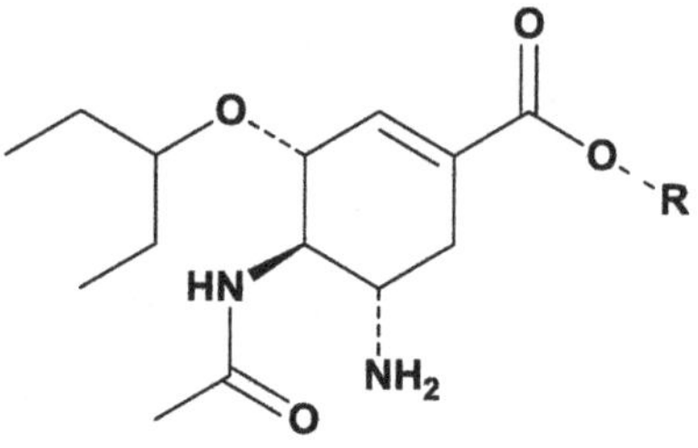

R = H; oseltamivir acid

R = $-C_2H_5$; oseltamivir

Pharmaceutical formulation is the most frequent method used to achieve slow release and prolong the duration of action of a given drug. However, a prodrug strategy can also be useful, as illustrated by the depot formulation of esters of steroid hormones. A conceptually different and particularly elegant approach to slow metabolic release has been achieved with bambuterol, a prodrug of the β_2-adrenoreceptor agonist terbutaline. Compared to terbutaline, bambuterol provides smooth and sustained plasma levels of terbutaline, and a greater symptomatic relief of asthma with a lower incidence of side effects.

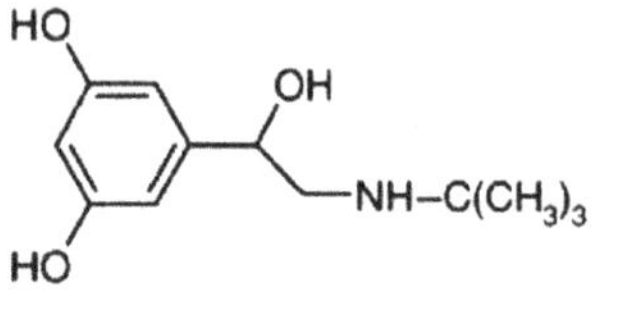

Terbutaline

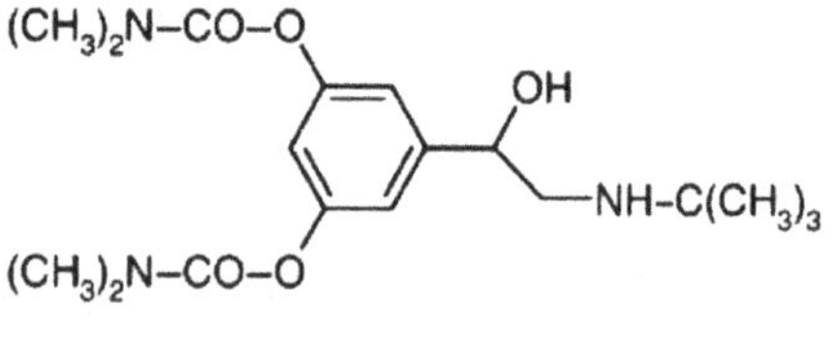

Bambuterol

One more pharmacokinetic objective of prodrugs is organ/tissue-selective delivery of a given drug. A very significant example is that of capecitabine, a multistep, orally active prodrug of 5-fluorouracil, an anticancer agent. Capecitabine is well absorbed orally and undergoes three activation steps resulting in high tumor concentrations of the active drug. It was first approved for the cotreatment of refractory metastatic breast cancer. Its therapeutic spectrum now includes metastatic colorectal cancer. Capecitabine thus affords an impressive gain in therapeutic benefit compared to 5-fluorouracil due to its oral bioavailability and a relatively selective activation in and delivery to tumors.

2.9.3 Pharmacodynamic Objectives

Pharmacodynamic objectives aim to decrease the systemic toxicity of a drug. This might be the masking of a reactive agent to improve its therapeutic index, or the *in situ* activation of a cytotoxic agent. The masking of a reactive agent to improve its therapeutic index is suitably illustrated by the successful anti-aggregating agent clopidogrel. This compound, whose molecular mechanism of action was poorly understood for years, is now known to be a prodrug. *In situ* activation to a cytotoxic agent is part of the well-known mechanism of action of the antibacterial and antiparasitic nitroarenes such as metronidazole (Testa, 2009).

Thus, it is clear that the objectives discussed above are interconnected. Thus, an improved solubility can greatly facilitate oral absorption, while improving the chemical stability of an active agent can allow tissue-selective delivery and even lead to its *in situ* activation. The objectives in the prodrug design can be summarized as indicated in the table below.

Pharmaceutical Objectives	Pharmacokinetic Objectives	Pharmacodynamic Objectives
• Improved solubility	• Improved oral absorption	• Masking of active agent to improve therapeutic index
• Improved chemical stability	• Decreased presystemic metabolism	• *In situ* activation of cytotoxic agent
• Improved taste, odor	• Improved absorption by nonoral route	
• Decreases irritation and pain	• Improved time profile	
	• Organ/tissue-selective delivery	

Although prodrug design is very challenging, it can still be more feasible and faster than searching for an entirely new therapeutically active agent with suitable ADME and toxicity properties. Another fallacy is that prodrugs are simply of academic interest and do not have industrial applications in attempts to overcome bioavailability and toxicity problems. A very good indication of the success of the prodrug approach can be obtained by examining the prevalence of prodrugs on the market. Currently about 10% of all globally marketed medicines can be classified as prodrugs, and in 2008 alone, 33% of all approved low-molecular-weight drugs were prodrugs. Despite these impressive numbers, only recently has the full potential of the prodrug approach begun to be appreciated in modern drug development, and many novel prodrug innovations await discovery.

The Rationale Behind the Use of Prodrug			
Barrier to Overcome	**Examples of Prodrugs**	**Preferable Site of Bioconversion**	**Common Functional Group Amenable to Prodrug Design**
Formulation and administration • Low aqueous solubility • Low shelf-life • Pain or irritation after local administration	Introducing ionizable or polar neutral group • Phosphates • Amino acid esters/amides • Sugar derivatives	During or after absorption • At brush border of enterocytes • In systemic circulation or after local administration • In systemic circulation by hydrolytic enzymes	–OH –SH –NH *(via spacer)* –COOH
Absorption • Poor membrane permeation • Low stability in GI tract • Substrate of efflux transporters	Masking polar ionized/unionized groups • Alkyl/aryl esters • Amino acid esters/amides	After absorption • In systemic circulation by hydrolytic enzymes	–OH –SH –NH –COOH $-OPO(OH)_2$
Distribution • Lack of site specificity • High degree of plasma protein binding	Targeting cell- or tissue-specific transporters • Amino acid esters/amides • Sugar derivatives	In the target tissue • By cell-specific hydrolytic enzymes or oxidoreductases	–OH –SH –NH –COOH
Metabolism and excretion • Short duration of action	Masking metabolically labile groups • Alkyl/aryl esters • Amino acid esters/amides	After absorption • In systemic circu-lation • In target tissue by hydrolytic enzyme	–OH –SH –NH –COOH $-OPO(OH)_2$
Toxicity • Lack of site specificity	Targeting tissue-specific enzymes or divergent condition of target tissue • Various different prodrug method	In the target tissue • By cell-specific enzymes • Because of altered pH or hypoxia	Depending on selected prodrug method (*usually bioprecursors*)
Life cycle management	Introducing any kind of promoiety • Various different prodrug methods	Depending on selected prodrug method	Depending on selected prodrug method (*usually carrier-linked prodrugs*)

2.10 EVALUATION OF PRODRUGS

Once the drug/pharmacological agent is converted into a prodrug, it is necessary to assess the new derivative (prodrug) by using several parameters. This is needed to check whether the synthesized prodrugs are fulfilling the need and criteria of the aforesaid purpose or not. This also gives direction to overcoming any unwanted properties of the existing drug. The evaluation of prodrug is also required to check the nature of promoiety and its interference in physicochemical property and activity as well. This confirms the safety of the drugs and thus increases the therapeutic effectiveness of the existing drug. The various evaluation parameters are as follows.

2.10.1 Physiochemical Parameters

The physicochemical parameters affecting drug action include solubility, lipophilicity, and partition coefficient. This is also applicable to prodrugs, because this is the first step to responsibly show the way for cleavage of prodrugs.

2.10.2 Pharmacokinetics Profile

As derivatization and synthetic methods are necessary to design a prodrug, of equal importance is its cleavage. The action of the drug is associated with the drug molecule only, and not with the prodrug. Chemical and enzymatic hydrolysis is the guiding step in the design of prodrugs. Unless there is hydrolysis by one or other means the parent drug will not be released. Thus, the pharmacokinetics profile and release pattern are vital for prodrugs.

2.10.3 Pharmacodynamics

As every drug must have adequate therapeutic activity, so too must every prodrug. To appraise the prodrug completely, the final step is to check the pharmacological activity. Evaluation of activity is done in order to ensure the retention or increase in the drug activity. In some cases, such as mutual prodrugs, where both drug and promoiety have activity, synergistic activity may result. All these activities concern the pharmacodynamic profile of the prodrug and are requirements in prodrug design.

REFERENCES

Albert, A., 1958. Chemical aspects of selective toxicity. Nature 182, 421–423.

Banerjee, P.K., Amidon, G.L., 1985. Design of prodrugs based on enzymes-substrate specificity. In: Bundgaard, H. (Ed.), Design of Prodrugs. Elsevier, New York, NY, pp. 93–133.

Han, H.K., 2000. Targeted prodrug design to optimize drug delivery. AAPS Pharm. Sci. 2 (1), 1–11.

Higuchi, T., Davis, S.S.I., 1970. Thermodynamic analysis of structure-activity relationships of drugs: prediction of optimal structure. J. Pharm. Sci. 59, 1376–1383.

Roche, E.B., 1977. Design of Biopharmaceutical Properties Through Prodrugs and Analogs. American Pharmaceutical Association, Washington, DC.

Sinkula, A.A., Yalkowsky, S.H., 1975. Rationale for design of biologically reversible drug derivatives: prodrugs. J. Pharm. Sci. 64, 181–210.

Stella, V., 1975. Pro-drugs: an overview and definition. In: Higuchi, T., Stella, V. (Eds.), Prodrugs as Novel Drug Delivery Systems. ACS Symposium Series. American Chemical Society, Washington, DC, pp. 1–115.

Stella, V.J., Charman, W.N., Naringrekar, V.H., 1985. Prodrugs. Do they have advantages in clinical practice? Drugs 29, 455–473.

Testa, B., 2009. Prodrugs; bridging pharmacodynamic/pharmacokinetic gaps. Curr. Opin. Chem. Biol. 13, 338–344.

CHAPTER 3

Types of Prodrugs

3.1 CLASSIFICATION OF PRODRUGS

A number of criteria are available to classify the various types and subtypes of prodrugs. While some of these criteria are quite useful in prodrug research, others are of a more historical and didactic interest (Testa, 2004).

The *criteria for classifying prodrugs* are:

1. Research-related criteria
2. Chemical criteria.

3.1.1 Research-Related Criteria

A historical discrimination can be made between deliberate and accidental prodrugs. The deliberately designed prodrugs are the result of chemical derivatization of an active drug molecule. Most prodrugs in clinical use are of this type, and they were developed to improve the pharmaceutical and/or pharmacokinetic properties of an active agent, be it a lead candidate, a lead compound, a clinical candidate, a drug candidate, or a marketed drug. A historically interesting example is that of hexamine. Hexamine is an intentional prodrug of formaldehyde used as a throat disinfectant for a long time.

$$\text{(hexamine)} + 6\,H_2O \xrightarrow{H^+} 4\,NH_3 + 6\,HCHO$$

Some prodrugs were serendipitous discoveries; the discovery of the active agent they generate will contribute significantly to understanding of their mechanism of action and may even lead to the discovery of a new therapeutic class. The most spectacular illustration of this was the discovery of the antibacterial medicine, prontosil, which has

Prodrug Design. DOI: http://dx.doi.org/10.1016/B978-0-12-803519-1.00003-9

been converted to its active metabolite sulfanilamide. This has been widely used since its discovery in 1935. The discovery of sulfa drugs was a milestone in medicinal chemistry and all sulfonamides using today are as a result of the discovery of prontosil.

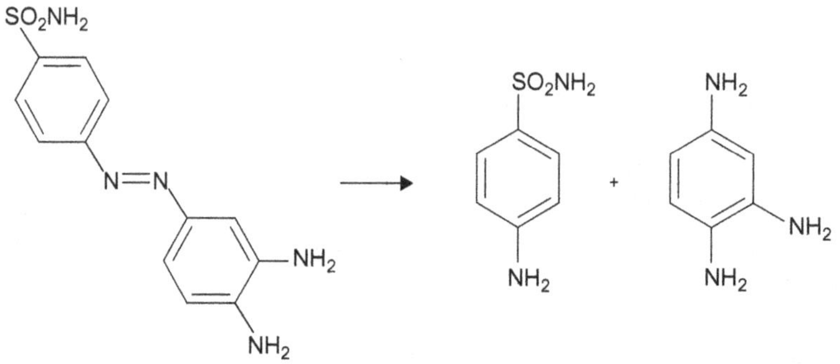

SO_2NH_2 N=N NH_2 NH_2 → SO_2NH_2 NH_2 + NH_2 NH_2 NH_2

3.1.2 Chemical Criteria

Another mode to classify the prodrugs is based on the chemical point of view. Based on chemical criteria prodrugs are sub-classified into different classes. This classification is widely used and so many prodrug derivatives are found to fall in this class/subclass:

1. Carrier-linked prodrugs
2. Mutual prodrugs
3. Bioprecursor prodrugs
4. Tripartate prodrugs
5. Polymeric prodrugs.

3.1.2.1 Carrier-Linked Prodrug

This is a compound that contains an active drug linked to a carrier group that can be removed enzymatically after hydrolysis. The bond to the carrier group must be liable enough to allow the active drug to be released efficiently *in vivo*. In carrier-linked prodrugs, a drug molecule, the usefulness of which is limited by its adverse physicochemical properties, is attached to a carrier group of promoiety to form a new compound, i.e., prodrug. Formation of prodrug does not alter the primary structure of the parent drug and the unique failure of this approach is that the physicochemical properties can be tailored by means of changing the structure of promoiety and intrinsic activity of the parent drug is assured through the *in vivo* cleavage of the prodrugs (Verma et al., 2009).

The carrier group must be nontoxic and biologically inactive when detached from the drug. Various adverse physicochemical properties of drug can be tailored and side effects can be minimized by attaching a nontoxic carrier group or promoiety to form a new compound, i.e., prodrug, from which the parent drug is regenerated *in vivo.*

A common example is dipivalyl ester of epinephrine, which enhances corneal absorption and inhibits the rapid metabolic destruction of epinephrine. In addition, prodrug produces less cardiovascular side effects.

Depending upon the nature of the carrier used, the carrier-linked prodrug may further be classified into:

1. *Cascade-latentiated prodrugs*
   A prodrug is further derivatized in a fashion such that only enzymatic conversion to prodrug is possible before the latter can cleave to release the active drug.
2. *Double prodrugs, pro-prodrugs*
   The double prodrug is a biologically inactive molecule which is transformed *in vivo* in two steps (enzymatically or chemically) to the active species.
3. *Macromolecular prodrugs*
   In these, macromolecules, such as polysaccharides, dextrans, cyclodextrins, proteins, peptides, and polymers, are used as carriers.
4. *Site-specific prodrugs*

The most important feature of an efficient drug is the correct site of action. It is necessary to deliver the drug precisely to the affected part of the body. In site-specific prodrugs, a carrier acts as a transporter of the active drug to a specific targeted site. The specific delivery can be obtained by tissue-specific activation of prodrug which results in metabolism by an enzyme that is either unique for the tissue or present at a higher concentration as compared to other tissues. Site-specific prodrugs are used for selective uptake systems, for urinary tract infections, or some anticancer chemotherapeutic agents.

3.1.2.2 Mutual Prodrug

A mutual prodrug consists of two pharmacologically active agents coupled together covalently so that each acts as a promoiety for the other and *vice versa.* Mutual prodrug design is really no different from the general drug discovery process, in which a unique substance is observed to have desirable pharmacological effects, and studies of its properties lead to the design of better drugs. The selected carrier may have the same biological action as that of the parent drug and thus might have a synergistic action, or the carrier may have some additional biological action that is lacking in the parent drug, thus ensuring additional benefit. The carrier may also be a drug that might help to target the parent drug to a specific site, organ, or cells, or may improve the site-specificity of a drug. The carrier drug may be useful to overcome some side effects of the parent drug as well (Bhosale et al., 2006).

The mutual prodrug concept is useful when two synergistic drugs need to be administered at the same site at the same time. Mutual prodrugs are synthesized toward a pharmacological objective of improving each drug's efficacy, optimizing delivery, and lowering toxicities.

The significant parameters which are to be considered before synthesis of a mutual prodrug are summarized below:

1. The candidate drugs selected for mutual prodrug synthesis can be from the same therapeutic category or from different therapeutic categories.
2. The candidates for making mutual prodrugs can be the pairs of drugs that are currently used in combination therapy in various therapeutic areas, provided each of those drugs possesses the requisite functional group(s).
3. The linkage between the first and second component should be a cleavable linkage preferably, under physiological conditions, such as those present in a mammalian body, particularly a human body.

Even if mutual prodrug design has proven highly beneficial in overcoming various undesirable properties of drugs, it can also give rise to a large number of newer difficulties, especially in the assessment of pharmacological, pharmacokinetic, toxicological, and clinical properties.

Benorylate, which is a common example of this category, is a prodrug of acetyl salicylic acid and paracetamol. A major advantage associated with this prodrug is in the treatment of chronic inflammation at a decreased dose and with a reduced risk of irritation.

Benorylate → Acetyle salicylic acid + Paracetamol

3.1.2.3 Bioprecursor Prodrug

The bioprecursor does not contain a temporary linkage between the active drug and carrier moiety, but is designed from a molecular modification of an active principle itself. It does not have any carrier or promoiety. Numerous drugs are currently known which exert pharmacological effects after their conversion into an active metabolite. Bioprecursor prodrugs are compounds that already contain the embryo of the active species within their structure, i.e., it contains latent functionality. A bioprecursor prodrug has a different structure that cannot be converted into active drug by simple cleavage of a group from the prodrug. This active species is liberated by metabolic or chemical transformation of prodrug.

The types of activation often involved are oxidative, reductive, or phosphorylation. Of these, oxidative is common, while phosphorylation is mainly used for the development of antiviral agents (Singhvi and Chaturvedi, 1997).

3.1.2.3.1 Oxidative Bioactivations

Nonsteroidal anti-inflammatory agents (NSAIDs) produce irritation of the stomach if taken for an extended period, this is due to the presence of an acidic group. Carboxylic functionalities are unionized in the highly acidic environment of the stomach. As a result, these agents are more lipophilic and may pass into the mucosa of the stomach. The intracellular pH of these cells is more basic than stomach lumen and NSAIDs become ionized resulting in back flow of H^+. This can be prevented if carboxylic functionalities are eliminated. *Nabumetone* contains no acidic functionality and passes through the stomach without producing irritation. Subsequent absorption occurs in the intestine and produces an active compound.

Series of oxidative
Decarboxylation

Active form of drug acts by inhibiting cyclooxygenase.

3.1.2.3.2 Reductive Bioactivations
This is only occasionally seen, it is less common than oxidative bioactivation due to the lower number of reducing enzymes. Mitomycin C, an antineoplastic agent (used in treating lung and bladder cancers), contains quinone, the functionality of which undergoes reduction resulting in hydroquinone. Quinone has an e^- withdrawing effect and hydroquinone has an e^- releasing effect. This allows electrons to participate in the expulsion of methoxide and, subsequently, the loss of carbamate to generate reactive species which may alkylate DNA.

3.1.2.4 Polymeric Prodrug

The conjugation of a drug with a polymer is called a polymeric prodrug. These are biocompatible (nontoxic, nonantigenic, nonteratogenic). Polymeric prodrugs are designed to improve the use of drugs for therapeutic applications. Successful bioconjugation depends upon the chemical structure, molecular weight, steric hindrance, and the reactivity of the biomolecule as well as the polymer. In order to synthesize a bioconjugate, both chemical entities need to possess reactive or functional groups, such as $-COOH$, $-OH$, $-SH$, or $-NH_2$. However, the presence of multiple reactive groups makes the task more complex.

The polymeric carrier can be either an inert or a biodegradable polymer. Selection of a suitable polymer and methodology to conjugate the same with different bioactive components is a critical step. The drug can be preset directly or via a spacer group onto the polymer backbone. The proper selection of this spacer opens the possibility of controlling the site and the rate of release of the active drug from the conjugate by hydrolytic or enzymatic cleavage (Hoste et al., 2004).

Three major types of polymeric prodrugs are currently being used:

1. Prodrugs that are broken down inside cells to form active substance(s).
2. Usually the combination of two or more substances (under specific intracellular conditions, these substances react to form an active drug).
3. Targeted drug delivery systems, usually includes three components: a targeting moiety, a carrier, and one or more active components.

An example is p–phenylene diamine mustard, which is covalently attached to polyamino polymer backbone polyglutamic acid.

Polymeric conjugates have several advantages over their low–molecular-weight precursors. The main advantages include: an increase in water solubility and therefore, enhancement of drug bioavailability; protection of drug from deactivation and preservation of its activity during circulation, and transport to the targeted organ or tissue; an improvement in pharmacokinetics; a reduction in antigenic activity; the ability to provide passive or active targeting of the drug specifically to the site of its action and may include several other active components that enhance the specific activity of the main drug (Khandare and Minko, 2006) (Figure 3.1).

3.1.2.5 Tripartate Prodrug

Structures of most prodrugs are bipartite in nature, whereby parent drug is attached directly to promoiety. Bipartite prodrug comprised of one carrier is attached to the drug. However, in some cases bipartite prodrug may be unstable due to the inherent nature of the drug–promoiety bond. This can be overcome by designing a tripartite prodrug, utilizing a spacer or connector group between the drug and promoiety. The spacer or connector group must be designed in such a way that the initial activation

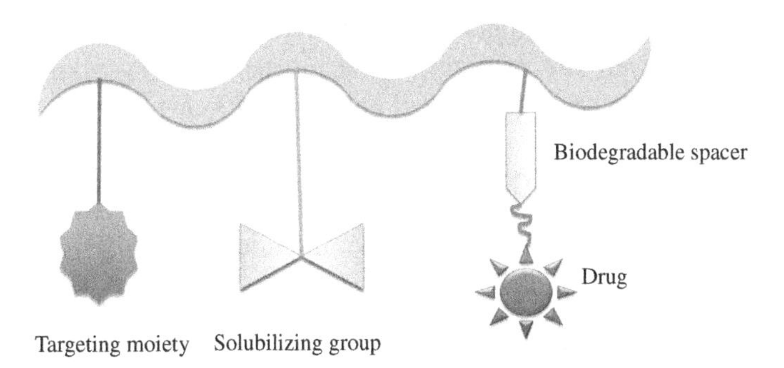

Figure 3.1 Ringsdorf's model of polymeric prodrug.

is followed by spontaneous cleavage of the remaining drug spacer bond under physiological conditions to release parent drug.

For example, a model tripartite prodrug P–(N–(tert-butyloxy carbonyl) lysyl amido benzyloxy carbonyl)–P–nitro aniline has been designed in which N–tert butyloxy carbomyl lysine group is the promoiety, P–amido benzyloxy carbonyl group is the spacer group and P–nitro aniline is the drug.

3.2 CRITERIA FOR PRODRUG

A well-designed carrier-linked prodrug should satisfy certain criteria. The linkage between the drug and the carrier should usually be a covalent bond. As a rule, the prodrug itself should be inactive or less active than the parent drug. The linkage should be bioreversible. The prodrug and the carrier released after *in vivo* enzymatic or nonenzymatic attack should be nontoxic. The generation of the active form must take place with rapid kinetics to ensure effective drug levels at the site of action. The bioavailability of carrier-linked prodrug is modulated by using a transient moiety. The lipophilicity is generally the subject of profound alteration of the parent molecule. The bioactivation process is exclusively hydrolytic and sometimes a redox system.

An ideal carrier should be without intrinsic toxicity. It should be nonimmunogenic and nonantigenic and should not accumulate in the body. It should possess a suitable number of functional groups for drug attachment and adequate loading capacity. It should be stable to chemical manipulation and autoclaving. It should be easy to characterize and should mask the liganded drug's activity until release of active agent at the desired site of action. In a mutual prodrug approach, the carrier should have some biological activity of its own.

3.3 CLASSIFYING PRODRUGS

Methods used to classify prodrugs have included:

- Therapeutic categories (e.g., anticancer prodrugs, antiviral prodrugs)
- Categories based on chemical linkages or moiety
- Functional categories based on strategic approaches to overcome particular deficiencies in the active drug, such as poor site specificity.

Classification of Prodrugs Based on Where They are Converted into Active Drugs				
Class	Conversion Site	Sub Type	Tissue Location of Conversion	Examples
I	Intracellular	I A	Therapeutic target	Zidovudine
			Tissues or cells	5-fluorouracil
		I B	Metabolic tissues (liver, lungs)	Captopril
				Cyclophosphamide
II	Extracellular	II A	Gastrointestinal fluid	Sulfasalazine
				Loperamide oxide
		II B	Systemic circulation	Fosphenytoin
				Bambuterol

While circumventing disadvantageous pharmacodynamic or pharmacokinetic properties is the primary goal of a prodrug, additional concerns are how fast and complete the prodrug converts into the active drug and whether it contributes significantly to the active drug's toxicity profile.

The Food and Drug Administration (FDA) has classified prodrugs into four categories based on their sites of conversion into the final active drug form. Type I prodrugs are those that are converted within a cell while type II prodrugs are converted outside the cell, predominantly in digestive or systemic fluids.

Both types I and II can be further categorized into subtypes based on whether the intracellular converting location is also the site of therapeutic action or whether the conversion occurs in the gastrointestinal fluids or systemic circulation. Because determination of drug actions has always been focused on the site and mechanism of action, classification of prodrugs based on cellular locations of conversion is in line with current FDA thinking. For example, if a prodrug is converted into an active drug in the gastrointestinal fluids, the safety and toxicity profile can be fully interpreted with existing knowledge of the active drug assuming that the conversion is complete, as validated by data showing no unconverted prodrug left at the gastrointestinal site and no measurable systemic prodrug.

From the standpoint of assessing the benefit and risk of a prodrug, a classification system based on the site of its conversion into the active drug form provides insight into the kinetics of the exchange process and the contributory role of both prodrug and active drug to the product's efficacy and safety (Wu, 2009).

However, the use of this classification system has limited the impact on the new chemical entity market exclusivity determination of a prodrug.

3.4 CHALLENGES AND LIMITATIONS IN PRODRUG DESIGN

An ideal prodrug should be stable to storage but must convert to the active drug under a specific set of conditions. These conditions for the activation of the prodrug will depend on its purpose and site of action. This strategy has various advantages. The biggest advantage is that the rate of conversion can be fine-tuned by selecting the structure of the promoieties. It is possible that the half-lives of these prodrugs might be different when studied *in vivo* as compared to the *in vitro* experiments. This will probably be as a result of the action of the esterases and other enzymes, but unlikely due to a change in the chemical rate of conversion. The studies of the structure of additional promoieties and how they affect the rate of conversion will provide additional diversity in prodrug chemistry for future use (Patil and Shirote, 2011).

The several challenges anticipated by medicinal chemists and biochemists carrying out prodrug research include additional work involved in synthesis, physicochemical profiling, pharmacokinetic profiling, and toxicological assessment. Two of these challenges are initiated, namely biological variability and toxicity potential (Jiunn and Anthony, 1997).

The problem in prodrug design arises due to biological diversity results from the huge number and evolutionary diversity of enzymes involved in xenobiotic metabolism. Inter- and intraspecies differences in the nature of these enzymes, as well as many other differences such as the nature and level of transporters, may render prodrug optimization difficult to predict and achieve. A chemical strategy developed by medicinal chemists to overcome the problem of biological variety is the development of prodrugs activated by non-enzymatic hydrolysis, for example, imines, Mannich bases, (2-oxo-l, 3-dioxol-4-yl) methyl esters, or oxazolidines. A more promising approach appears to be the two-step activation of carrier-linked prodrugs, involving first a relatively facile enzymatic hydrolysis to unmask a nucleophilic group, followed by a nonenzymatic, intramolecular nucleophilic substitution and cyclization.

The second problem associated with prodrug design is its toxicity. The toxicity potential of some prodrugs is a serious limitation. Schematically, prodrug designers must be aware of at least two specific sources of toxicity, namely the formation of unexpected metabolite from the total drug conjugates, or it may be due to an inert carrier generated by cleavage of promoiety and drug conjugate which is converted into toxic metabolite. The former case is illustrated by the liberation of formaldehyde, as seen with Mannich bases or some double esters. In some cases, the prodrug might consume a vital cell constituent, such as glutathione, during its activation stage and thus lead to toxicity (Testa and Joachim, 2003).

REFERENCES

Bhosale, D., Bharambe, S., Gairola, N., Dhaneshwar, S.S., 2006. Mutual prodrug concept: fundamentals and applications. Indian J. Pharm. Sci. 68, 286–294.

Hoste, K., Winne, K.D., Schacht, E., 2004. Polymeric prodrugs. Int. J. Pharm. 277, 119–131.

Jiunn, H.L., Anthony, Y.H., 1997. Role of pharmacokinetics and metabolism in drug discovery and development. Pharmacol. Rev. 49 (4), 404–449.

Khandare, J., Minko, T., 2006. Polymer–drug conjugates: progress in polymeric prodrugs. Prog. Polym. Sci. 31, 359–397.

Patil, S.J., Shirote, P.J., 2011. Prodrug approach: an effective solution to overcome side-effects. Int. J. Med. Pharm. Sci. 1 (7), 1–13.

Singhvi, I., Chaturvedi, S.C., 1997. Prodrugs. East. Pharm., 57–59.

Testa B., Joachim M.M., Metabolic Hydrolysis and Prodrug Design, Hydrolysis in Drug and Prodrug Metabolism: Chemistry, Biochemistry, and Enzymology, John Wiley & Sons Canada Ltd; 2003; pp 800.

Testa, B., 2004. Prodrug research: futile or fertile. Biochem. Pharmacol. 68, 2097–2106.

Verma, A., Verma, B., Prajapati, S., Tripathi, K., 2009. Prodrug as a chemical delivery system: a review. Asian J. Res. Chem. 2 (2), 100–103.

Wu, K.-M., 2009. A new classification of prodrugs: regulatory perspectives. Pharmaceuticals 2, 77–81.

CHAPTER 4

Approaches for Prodrugs

The several approaches for designing of prodrugs are the chemical methods of derivatization of parent drug. The drugs that haverestricted use for various reasons are modified into prodrugs. The chemical derivatization of a drug is dependent upon the nature of the functional group. Attachment of promoiety is possible only when there is compatibility between drug and promoiety. The success rate of the resultant prodrug is linked to the nature and binding of drug with the promoiety. Moreover, equally important is the cleavage of drug and promoiety. This also, in turn, depends upon the chemical nature and functional group.

4.1 PROMOIETY

A functional group used to modify the structure of pharmacologically active agents (parent drug) to improve physicochemical, biopharmaceutical, or pharmacokinetic properties is called promoiety. It should ideally be safe and rapidly excreted from the body. The choice of promoiety should be considered with respect to the disease state, dose, and the duration of therapy. The design of prodrug structure depends on the nature of the parent drug as well as promoiety.

Some of the most common functional groups that are amenable to prodrug design include carboxylic, hydroxyl, amine, phosphate/phosphonate, and carbonyl groups. Prodrugs typically produced through the modification of these groups include esters, carbonates, carbamates, amides, phosphates, and oximes. Drugs containing hydroxyl groups, alcohols, and phenols can have a variety of physical/chemical properties that have advantages and disadvantages. Many hydroxyl groups often impart polar properties to molecules and also provide a handle for prodrug intervention whereby the properties of the parent drug may be manipulated (Dhaneshwar and Stella, 2007).

Prodrug Design. DOI: http://dx.doi.org/10.1016/B978-0-12-803519-1.00004-0

However, other uncommon functional groups have also been investigated as potentially useful structures in prodrug design. For example, thiols react in a similar manner to alcohols and can be derivatized to thioethers and thioesters. Amines may be derivatized into imines and N-Mannich bases. The prodrug structures for the most common functionalities are illustrated in Figure 4.1.

Esterification of the hydroxyl group(s) has been one of the preferred prodrug strategies. By appropriate esterification of the hydroxyl group(s), it is possible to obtain derivatives with desirable lipophilicity as well as *in vivo* liability; the latter is influenced by electronic and steric factors. Although both carboxylic acids and alcohols can be derivatized to their ester derivatives, the esters of carboxylic acids have more clinical success. The synthesis of an ester prodrug is often straightforward. Once in the body, the ester bond is readily hydrolyzed by ubiquitous esterases found in the blood, liver, and other organs and tissues. Phosphate ester prodrugs are typically designed for hydroxyl and amine functionalities of poorly water-soluble drugs with the aim of enhancing their aqueous solubility to allow more favorable oral or parenteral administration. Prodrugs can be formed as other derivatives such as amides, imides, urides; several ring formation derivatives, glycolamide esters, and carbamates.

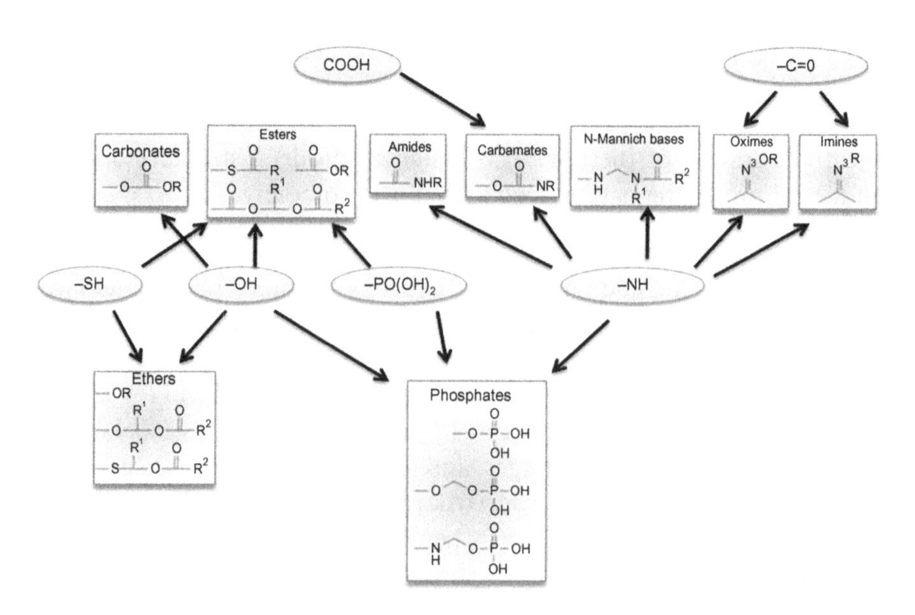

Figure 4.1 Prodrugs for compounds containing different functionalities.

The general structure representing the chemical nature of prodrug is:

$$\mathbf{DRUG - X - R}$$

where **X** is the functional group in a drug, in which promoiety **R** can be attached.

$$\mathbf{X}\,\text{can be} - COOH, -NH, -OH, -SH, -CONH, -SO_2NH, -C{=}O$$

The overall nature of prodrug should be inactive and nontoxic, easily synthesizable, chemically stable outside the site of action, and bioreversible (parent drug must be regenerated *in vivo*).

4.2 FUNCTIONAL GROUPS COMPLIANT FOR DESIGN OF PRODRUG

Ideally, the design of an appropriate prodrug structure should be considered at the early stages of preclinical development, bearing in mind that prodrugs might alter the tissue distribution, efficacy, and the toxicity of the parent drug. Several important factors should be carefully examined when designing a prodrug structure, including:

- *Nature of parent drug*:
  Selection of functional group which needs to be derivatized and modify chemically.
- *Nature of promoiety*:
  This is the most important part of prodrug design. Ideally the prodrug should be safe and rapidly excreted from the body. The choice of promoiety should be considered with respect to the disease state, dose, and duration of therapy. As single parent drug can be derivatized to several prodrugs, this change is due only to the change in nature of promoiety. This difference in the promoiety varies the cleavage of drug and promoiety bond.
- *Nature of parent and prodrug*:
  The pharmacokinetic properties, such asabsorption, distribution, metabolism, and excretion (ADME), related to the chemical nature of the drug and promoiety and its cleavage need to be comprehensively understood.
- *Degradation by products*:
  Formation of by-products may create difficulty designing the prodrug. Degradation is possible due to susceptibility of the functional group undergoing breakdown under various physiological

conditions. These can affect chemical and physical stability and lead to the formation of new degradation products.

4.3 BIOREVERSIBLE DERIVATIVES FOR VARIOUS FUNCTIONAL GROUPS

Various types of functional groups are present in different therapeutic agents. These functional groups react with other functional groups of nontoxic promoiety to form prodrugs. Some of the most common functional groups that are amenable to prodrug design include carboxylic, hydroxyl, amine, phosphate/phosphonate, and carbonyl groups. Prodrugs typically produced via the modification of these groups include esters, carbonates, carbamates, amides, phosphates, and oximes. However, other uncommon functional groups have also been investigated as potentially useful structures in prodrug design. For example, thiols react in a similar manner to alcohols and can be derivatized to thioethers and thioesters. Amines may be derivatized into imines and N-Mannich bases. Various prodrugs for compounds containing different functional groups are given in Figure 4.1 (Rautio et al., 2008).

Ester derivatives are suitable prodrugs for therapeutic agents containing carboxyl and hydroxyl functional groups. This is the most commonly seen type of prodrug. Ester prodrugs are most often used to enhance the lipophilicity, and thus the passive membrane permeability, of water-soluble drugs by masking charged groups such as carboxylic acids and phosphates. An additional factor that contributes to the popularity of esters as prodrugs is that the synthesis of an ester prodrug is often straightforward. Chemical reactivity of esters is readily predictable on the basis of the steric and electronic properties of the substitutes in both the acyl and alcohol molecules and on the other hand hydrophilic properties and charge of ester may play a major role in enzyme hydrolysis.

Once in the body, the ester bond is readily hydrolyzed by ubiquitous esterases found in the blood, liver, and other organs and tissues, including carboxylesterases, acetylcholinesterases, butyrylcholinesterases, paraoxonases, and arylesterases. In addition to these agents, microflora present within the gut produce a wide variety of enzymes capable of hydrolyzing esters.

However, one significant challenge with ester prodrugs is the accurate prediction of pharmacokinetic disposition in humans, owing to significant differences in specific carboxylesterase activities in preclinical species.

$$\text{Drug—C(=O)—O—Promoiety} \;\text{ or }\; \text{Drug—O—C(=O)—Promoiety} \xrightarrow{\text{Esterase}} \text{Drug—C(=O)—OH + HO Promoiety} \;\text{ or }\; \text{Drug—OH + HO—C(=O)—Promoiety}$$

Several alkyl and aryl ester prodrugs are in clinical use, of which angiotensin-converting enzyme (ACE) inhibitors are some of the most successful. However, the relatively slow and incomplete bioconversion of some simple alkyl esters in human blood can sometimes result in lower than predicted bioavailability. In some cases, faster bioactivation has been achieved by the use of a double prodrug (pro-prodrug), which requires an enzymatic breakdown, after which a spontaneous chemical reaction releases the parent drug. The double prodrug approach has been the preferred choice when preparing oral acyloxyalkyl ester prodrugs of β-lactam antibiotics.

4.4 PHOSPHATE ESTERS AS PRODRUGS OF HYDROXYL OR AMINE FUNCTIONALITIES

Phosphate ester prodrugs are typically designed for hydroxyl and amine functionalities of poorly water-soluble drugs. This design of prodrug enhances the aqueous solubility to allow more favorable oral or parenteral administration. The synthesis of phosphate prodrugs is somewhat straightforward, and the presence of the dianionic phosphate promoiety usually raises the aqueous solubility. Phosphate prodrugs typically display excellent or adequate chemical stability and rapid bioconversion to parent drug by phosphatases present at the intestinal brush border or in the liver. Unlike carboxylic acid esters, phosphate esters are typically hydrolyzed at similar rates in different preclinical species by alkaline phosphatases. Even though several phosphate prodrugs are reported, only a few have crossed the development phase and been used for orally. This is due to the challenges that occurred in the design (Delgado and Remers, 1998).

4.5 AMIDES AS PRODRUGS OF CARBOXYLIC ACIDS AND AMINES

Amides are derivatives of amines and carboxylic acid functionalities. Derivatization to amides has not been used as a prodrug strategy owing to their relatively high enzymatic stability *in vivo*. Also, lack of amidase enzymes necessary for hydrolysis might be another reason. An amide bond is usually hydrolyzed by ubiquitous carboxylesterases, peptidases, or proteases. Amides are often designed for enhanced oral absorption by synthesizing substrates of specific intestinal uptake transporters.

4.6 PRODRUG FOR AMIDES, IMIDES, AND OTHER ACIDIC COMPOUNDS

4.6.1 N-Mannich Bases and Acyloxy Derivatives

N-Mannich bases can function as a prodrug candidate for compounds such as amides, imides, and urea derivatives. Similarly, N-α-acyloxy alkylation of various amides, imides, and N-heterocyclic amines also were adopted as a common approach to obtain prodrugs. Though the derivatives showed good stability in aqueous solution *in vitro*, they are in general rapidly cleaved *in vivo* by virtue of enzyme-mediated hydrolysis.

$$R_1{-}C(=O){-}NH{-}CH_2{-}HN^{+}(R_2)(R_3) \rightleftharpoons R_1{-}C(=O){-}NH{-}CH_2{-}N(R_2)(R_3) + H^{+}$$

$$R_1{-}C(=O){-}NH{-}CH_2{-}N(R_2)(R_3) \longrightarrow R_1{-}C(=O){-}NH + H_2C{=}N(R_2)(R_3)$$

$$R_1{-}C(=O){-}NH \xrightarrow{H^{+}} R_1{-}C(=O){-}NH_2$$

$$H_2C{=}N(R_2)(R_3) \xrightarrow{OH^{-}} HO{-}CH_2{=}N(R_2)(R_3) \longrightarrow H{-}C(=O){-}H + {}^{+}NH(R_2)(R_3)$$

Reaction mechanism showing decomposition of Mannich base

4.6.2 N-Acyl Derivatives

Plasma enzyme-catalyzed hydrolysis of the N-acyl derivatives makes N-acylation of amide or imide fruitful in some cases, such as N-acetyl-5-fluorouracil and N-ethoxy carbonyl-2-fluorouracil. Improved physicochemical properties and easy bioconversion of N-acyl derivative of 5-fluorouracil enhances the oral and rectal absorption of the parent drug.

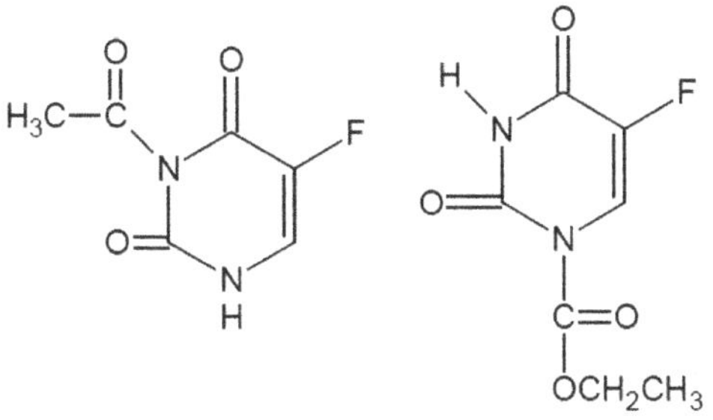

N-Acyl derivative and N-ethoxycarbonyl derivative of 5-Fluorouracil

4.6.3 N-Hydroxy Methyl Derivatives

The N-hydroxyl methyl derivatives of amides or imide-type compounds are more water-soluble than the parent compounds. By replacing a proton bind to nitrogen atom by a hydroxyl methyl group, intra- or intermolecular hydrogen bonding in such molecules may be increased, resulting in a decrease in melting point and an increase in water solubility.

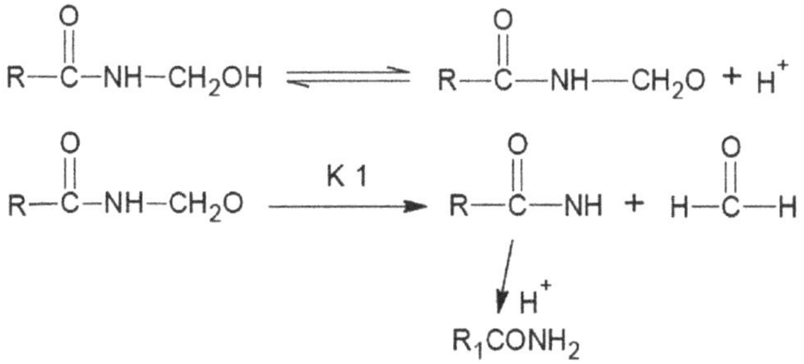

$$R-\overset{O}{\overset{\|}{C}}-NH-CH_2OH \rightleftharpoons R-\overset{O}{\overset{\|}{C}}-NH-CH_2O + H^+$$

$$R-\overset{O}{\overset{\|}{C}}-NH-CH_2O \xrightarrow{K1} R-\overset{O}{\overset{\|}{C}}-NH + H-\overset{O}{\overset{\|}{C}}-H$$

$$\downarrow H^+$$

$$R_1CONH_2$$

4.7 PRODRUGS FOR AMINES

Prodrugs of amines are generally designed by making their amide, N-(acyloxy alkoxy carbonyl) derivatives and oxazolidine derivatives.

4.7.1 N-(Acyloxy Alkoxy Carbonyl) Derivatives and Amide Derivatives

The utility of the N-(acyloxy alkoxy carbonyl) derivative is limited due to its resistance to undergo enzymatic cleave *in vivo*. However, certain activated amides are chemically labile and also certain amides formed with amino acids may undergo enzymatic cleavage. For example, the γ-glutamyl derivatives of dopamine, L-dopa and sulfamethoxazole are rapidly hydrolyzed by γ-glutamyl trans peptidase *in vivo.* Similarly N-glycyl derivative, midorin, and N-1-isoleucine derivative of dopamine are the enzymatically labile amide prodrugs.

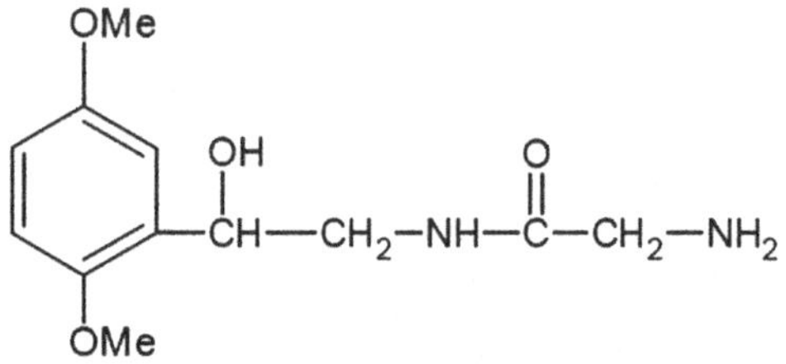

OMe
OH O
CH—CH₂—NH—C—CH₂—NH₂
OMe

4.7.2 Oxazolidines

Oxazolidines are cyclic condensation products of β-amino alcohols and aldehydes or ketone, and they undergo a facile and complete hydrolysis in aqueous solution. Alterations in carbonyl moiety control the rate of formation of a given β-amino alcohol. Oxazolidines are weaker bases (pKa 6–7) than parent β-amino alcohols and found to be more lipophilic than the parent compound at physiological pH. For example, the oxazolidine prodrug of phenylephrine prepared from pivaldehyde has penetrated the cornea much more easily than the parent drug as a result of increased lipophilicity.

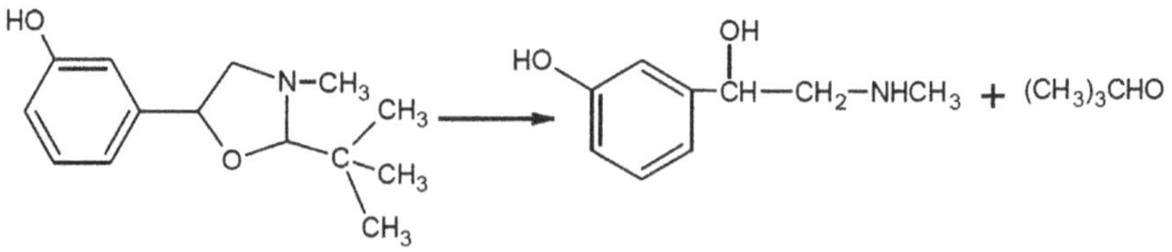

HO
N–CH_3
CH_3
O C CH_3
CH_3
HO OH
CH—CH_2—$NHCH_3$ + $(CH_3)_3CHO$

4.8 PRODRUGS WITH CARBONYL GROUPS

The weakly basic character of carbonyl-containing drugs may be advantageous as the transformation of such drugs into oxazolidine introduces a readily ionizable moiety, which allows the preparation of derivatives with increased aqueous solubilities at acidic pH.

4.8.1 Thiazolidines

Thiazolidines have been applied as prodrug derivatives for various steroids containing a 3-carbonyl group to improve their topical anti-inflammatory activity. Thiazolidine derivatives of hydrocortisone and hydrocortisone 21-acetate prepared with cysteine esters to related β-aminothiols have been shown to be readily converted to the parent corticosteroids at conditions similar to those prevailing in the skin, thus meeting the requirement for a prodrug.

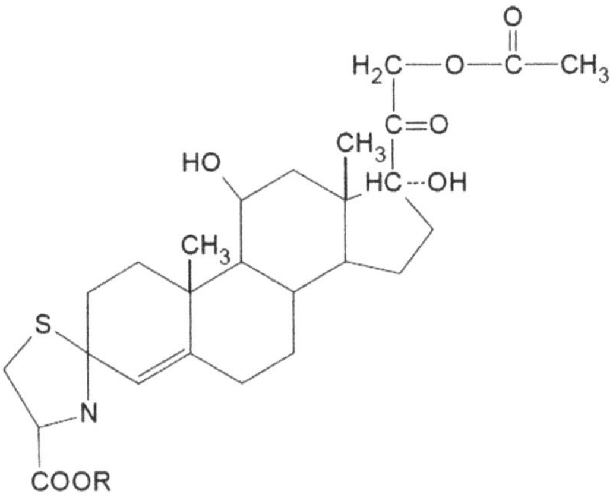

4.8.2 Enol Esters

The enol form of keto–enol equilibrium under proper conditions can be trapped by alkylation or acylation. Such enol esters and ethers may readily undergo hydrolysis with liberation of free enol, which then reverts to the keto form almost instantaneously. In the presence of plasma or liver enzymes, the enol esters are readily hydrolyzed. For example, the chemical stability of enol ester of acetophen is similar to that of phenol ester with maximum stability at pH 3.3. In contrast it is rapidly hydrolyzable in plasma and liver enzymes.

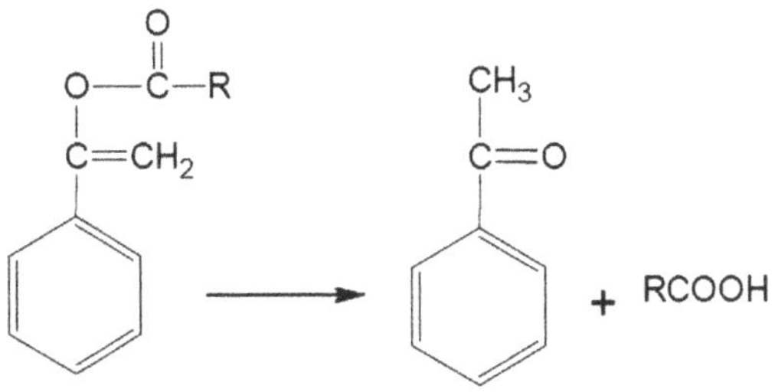

$$\text{C}_6\text{H}_5\text{–C(=CH}_2\text{)–O–C(=O)–R} \longrightarrow \text{C}_6\text{H}_5\text{–C(=O)–CH}_3 + \text{RCOOH}$$

4.9 TYPES OF PROMOIETIES USED IN DESIGNING OF PRODRUGS

As shown in Figure 4.2, a wide variety of promoieties have been used to overcome liabilities associated with drugs. The selection of promoiety depends on the purpose of the prodrug, type of functional groups

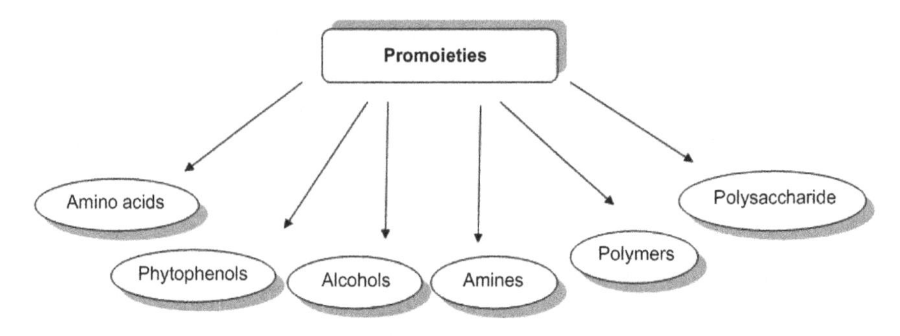

Figure 4.2 Types of promoieties used in designing of prodrugs.

available on the parent drug, chemical and enzymatic conversion mechanisms of prodrug to parent drug, safety of the promoiety, and ease of manufacturing (Dhokchawle et al., 2014).

4.9.1 Amino Acids

The coupling of drugs with amino acids chemically conjugates to form amide linkage. Conjugation with amino acid proves to be safer promoiety and serves various advantages (Vig et al., 2013).

1. They are normal dietary constituents and are nontoxic in moderate doses.
2. They have gastroprotective action.
3. Large structural diversity and the physicochemical properties of drug can be changed based on the nature of the amino acid selected for conjugation.
4. Drug targeting can be achieved by proper selection of amino acids.
5. Few amino acids have intrinsic anti-inflammatory activity.
6. Well-established prodrug chemistry.
7. Commercial availability.

Amino acid prodrugs were designed such that they have reasonably good chemical stability and will convert to parent drug. The high chemical stability reduced the conversion of the prodrug to the parent drug during the shelf life and normal handling of the prodrug, while activation by enzymes allows for rapid and quantitative conversion to the parent drug. The amide bond is less polarized than the ester bond due to the lower electronegativity of the nitrogen compared to the corresponding oxygen, thus rendering the amide bond more resistant than the ester bond to chemical hydrolysis. By selection of the proper amino acid, polarity, solubility profile, and acid base properties of a given drug molecule can be altered. The following table shows a few examples of prodrugs that are reported using amino acids as promoieties and the advantage achieved thereof.

Sr. No.	Amino Acids Used as Promoieties	Conjugation with Drug	Advantage
1	Glycine, phenylalanine, tryptophan, L-valine, L-isoleucine, L-alanine, L-glutamic acid, L-aspartic acid	Ketorolac	Controlled release and decreased gastrointestinal side effects
2	L-Alanine and histidine	Aceclofenac	Reduction in gastrointestinal side effects
3	L-Tryptophan, histidine, DL-alanine, L-phenylalanine	Flurbiprofen	Reduction in gastrointestinal side effects
4	Phenylalanine, valine, and proline	Floxuridine	Improved solution stability and fast enzymatic conversion rates
5	Glycine, alanine, lysine, leucine, phenylalanine	Dapsone	Improved water solubility
6	Valine	Saquinavir, indinavir, and nelfinavir	Improved pharmacological and pharmacokinetic profile
7	Glycine, cysteine, phenylalanine, tryptophan, and lysine	Cinmetacin	Reduction in gastrointestinal side effects

4.9.2 Polysaccharides

Polysaccharides are used as promoieties specifically for colon-targeting drug delivery. Various polysaccharides, such as cyclodextrin, dextran, pectin, chitosan, and chondroitin are conjugated with drugs. Chondroitin sulfate, a copolymer of D-glucuronic acid and sulfated N acetyl D-galactosamine, is an important structural component in connective tissue and cartilage. It can be used as a good candidate for colon-targeted drug carriers. Cyclodextrins belong to the family of oligosaccharides obtained by enzymatic degradation of starch. They are nontoxic and thought to be one of the most suitable promoieties to reduce ulcerogenic tendency. Dextran has excellent physicochemical properties and physiological response, and a unique pharmacokinetic profile. It was investigated as a macromolecular carrier for delivering drugs and also demonstrated that it is useful to target therapeutic agents to the liver. The following table gives an elaborative application of the uses of polysaccharides with their advantages (Praveen et al., 2009).

Sr. No.	Polysaccharide Used as Promoieties	Conjugation with Drug	Advantage
1	Chondrotin sulfate	Ibuprofen, ketoprofen and naproxen	Synergistic anti-inflammatory effect and delayed release of drug
2	Cyclodextrin	Mefenamic acid	Reduction in ulcerogenic potential
3	Chitosan glucosamine	Metronidazole	Colon-targeted drug delivery
4	Dextran	Flurbiprofen and suprofen	Improved analgesic and antipyretic effect with low ulcerogenic potential
5	Dextran	Metaxalone	Improved pharmacokinetics and longer half-life
6	Dextran conjugates	Valproic acid	Reduction in hepatotoxicity and ulcerogenicity
7	Dextran conjugates	Acyclovir	Targeting of antiviral to liver

4.9.3 Alcohols

Esters have dominated research because they have ideal characteristics, exhibiting (reasonable) sufficient chemical stability *in vitro* and because of their ability to function as esterase substrates for *in vivo* regeneration. Cyclohexanol, cyclopentyl alcohol, isobutyl alcohol, and t-butyl alcohol are alcohols used as promoieties in synthesis of ester prodrugs. Iodomethyl pivalate and 2-bromo ethyl acetate are also used. A variety of nonsteroidal anti-inflammatory agents bearing the carboxylic acid group are coupled with alcohols to afford ester. This leads to a reduction in the gastrointestinal side effects of drugs.

4.9.4 Phytophenols

Phytophenols are used traditionally for their medicinal as well as their flavoring properties, with well-documented safety profiles. Phytophenols are used as carriers for prodrugs in an attempt to combine anti-inflammatory and antioxidant properties. Naturally occurring phenolic antioxidants are thymol, guiacol, and eugenol, whereas menthol is n alcoholic compound. Menthol, thymol, eugenol, guiacol, vanillin, and umbelliferone are the promoieties used in prodrugs. The following table provides details of the advantages of coupling with phytophenols. These might be considered as mutual prodrugs due to pharmacological activity, especially the analgesic activity associated with these phytophenols (Redasani and Bari, 2012).

Sr. No.	Phytophenols Used as Promoieties	Conjugation with Drug	Advantage
1	Menthol, thymol, eugenol	Ibuprofen	Synergistic analgesic action and decreased gastrointestinal toxicity
2	Guiacol, eugenol, thymol, vanillin, umbelliferone	Diclofenac	Reduced ulcerogenic side effects
3	Guiacol	Mefenamic acid	Synergistic anti-inflammatory activity and less toxicity

4.9.5 Amines

Several compounds, such as propylamine, diethylamine, cyclohexyl amine, 2-amino ethyl amine, 2-hydroxyl ethyl amine, ethylenediamine, benzathine, and cysteamine are used as promoieties in the synthesis of amide prodrugs. These form amide bonds with carboxylic groups of drug moiety. Examples are quoted in the following table.

Sr. No.	Amine Used as Promoieties	Conjugation with Drug	Advantage
1	Propylamine, diethylamine, cyclohexyl amine, 2-amino ethyl amine, 2-hydroxyl ethyl amine	Ketoprofen	Reduced gastrointestinal side effects, improved analgesic activity
2	Ethylenediamine and benzathine conjugate	Ibuprofen	Reduced gastrointestinal side effects, improved analgesic activity
3	Heterocyclic amide	Ibuprofen	Improved analgesic activity, lower ulcerogenic activity
4	Amide derivatives	Diclofenac	Lower ulcerogenic activity
5	Cysteamine	Ibuprofen and indometacin	Antioxidant activity and lower ulcerogenic activity
6	Glycine amides	Ketoprofen	Lower ulcerogenic activity

4.9.6 Polymers

The delivery of biomolecules using polymeric materials has attracted considerable attention from polymer chemists, chemical engineers, and pharmaceutical scientists. The conjugation of a biologically active compound with a polymer is one of the many methods for altering and controlling the pharmacokinetics, biodistribution, and often the toxicity of various drugs. The task of obtaining a versatile polymer as an ideal candidate in drug delivery can be intricate as it has to surmount several vigorous clinical barriers. While designing the polymeric conjugates considerable attention has to be given to selection of the proper drug and polymer candidate.

4.9.6.1 Requirements for Selecting Polymers as Candidate Drug Carriers

- Availability of suitable functional groups –COOH, –OH, –SH, or $-NH_2$ for covalent coupling with drugs
- Biocompatibility: preferably nontoxic, nonimmunogenic
- Biodegradability or a molecular weight below the renal excretion limit
- Availability: reproducibly manufactured and conveniently administered to patients
- Water solubility: hydrophilic to ensure water solubility
- Low polydispersity, to ensure an acceptable homogeneity of the final conjugates.

4.9.6.2 Classification of Polymers Used for Bioconjugation

Based on their origin polymers used for bioconjugation polymers are classified as either synthetic or natural.

Synthetic polymers can be widely used because the properties of these molecules can be modified by varying their structures. The commonly used polymers of this class are:

Polyethylene glycol (PEG) is a polyether compound with many applications, from industrial manufacturing to medicine. PEG is particularly attractive as it is used as a pharmaceutical excipient and is known to be nontoxic and nonimmunogenic. It has a flexible, highly water-soluble chain that extends to give a hydrodynamic radius of effective range. Its high degree of hydration means the polymer chain effectively has a "water shell," and this helps to mask the drug to which it is bound.

Vinyl polymers are prepared by radical polymerization of the corresponding vinyl monomer. This copolymerization results in the formation of varied polymers with different polymer properties. Examples of this copolymerization are molecules like N-(2-hydroxypropyl) methacrylamide (HPMA), poly (styrene-co-maleic acid/anhydride) (SMA). Vinyl polymers are nonimmunogenic and nontoxic, they reside well in the blood circulation, and are frequently used as macromolecular carriers for low-molecular-weight drugs to enhance therapeutic efficacy and limit side effects. This prodrug, by means of SMA, has been successfully used for the treatment of hepatocellular carcinoma.

Divinyl ether (DVE) and *maleic anhydride* (MA) copolymerize with radical and show a wide variety of biological activities. They have antitumor activity; they induce the formation of interferon; they have antiviral, antibacterial, and antifungal activity; they also have an anticoagulant and an anti-inflammatory agent.
Polyethylenimine (PEI) is a polymer with a repeating unit composed of the amine group and two carbon spacers. Linear PEIs contain all secondary amines. It is used to overcome the nuclear barrier and yields the highest transfection rates.

The commonly used natural polymers are:

Dextran is a complex, branched glucan (polysaccharide made of many glucose molecules) composed of chains of varying lengths ranging from 3 to 2000 kDa. It is biocompatible and biodegradable. It is biologically active and possesses thrombolytic activity and is nonimmunogenic and nontoxic.
Chitosan is a linear polysaccharide composed of randomly distributed β-(1-4)-linked D-glucosamine (deacetylated unit) and N-acetyl-D-glucosamine (acetylated unit). It enhances the transport of polar drugs across epithelial surfaces, and is biocompatible and biodegradable. It helps in natural blood clotting and blocks nerve endings and hence reduces pain.
Pullulan is a polysaccharide polymer consisting of maltotriose units, also known as α-1, 4-α-1, 6-glucan. It is biodegradable, with low immunogenicity and poly-functionality, having fair solubility in aqueous and a few organic solvents, and is blood-compatible, nontoxic, nonmutagenic, and noncarcinogenic.

Polymeric prodrugs of several drugs have been synthesized and evaluated as shown in the following table.

Sr. No.	**Polymer Used as Promoieties**	**Conjugation with Drug**	**Advantage**
1	PEG	Theophylline	Improved biopharmaceutical properties
2	PEG	Ketoprofen	Extended pharmacological effect due to delayed release
3	2-Hydroxyl methyl acrylate	Naproxen	Enhanced potency and longer duration of action
4	PEG 2000	Warfarin	Improved biopharmaceutical properties
5	PEG 5000, 10000	Metronidazole	Improved pharmacokinetic properties
6	PEG esters	Methotrexate	Improved stability and drug delivery

7	PEG	Ibuprofen	Extended duration of action
8	PEG	Theophylline	Improved release of parent drug
9	PEG	Mesalazine	Colon-specific drug delivery
10	Acrylates	Ibuprofen	Increased anti-inflammatory activity

4.10 COUPLING OF DRUG AND POLYMER THROUGH SPACERS

Spacers may be incorporated during bioconjugation to decrease the crowding effect and steric hindrance. Spacers can enhance ligand protein binding and have application in prodrug conjugates. Amino acids, such as glycine, alanine, and small peptides are widely used as spacers due to their chemical versatility for covalent conjugation and biodegradability. The α-amino acids in peptides and proteins consist of a carboxylic acid and an amino functional group attached to the same tetrahedral carbon atom which extends diversity for conjugation with hydroxyl, carboxyl, or amino groups of polymers or biomolecules. Moreover, amino-acid-based spacers are short-chained, reactive, and biocompatible and may release the active agent from the conjugate. Difunctional amino acids such as 6-amino caproic acid or 4-amino butyric acids have been used as spacer arms between the polymers and the ligands for applications in biotechnology.

A polymeric prodrug of alkylating agent mitomycin C with poly [N5-(2-hydroxyethyl)-L-glutamine] (PHEG) using oligopeptide spacers was designed predominantly for enzymatic degradation which released parent drug in a rate-dependent manner. Hetero bifunctional coupling agents containing succinimidyl group have also been used extensively as spacers (Rohini et al., 2013).

REFERENCES

Delgado J.N., Remers W.A., 1998. Drug latentiation and prodrugs. Wilson and Gisvold's Text book of Organic, Medicinal and Pharmaceutical Chemistry, tenth ed., Lippincott Williams and Wilkins Philadelphia, pp. 123–138.

Dhaneshwar, S.S., Stella, V.J., 2007. Prodrugs of alcohols and phenols. Prodrugs Biotechnol. Pharm. Aspects V, 731–799.

Dhokchawle, B.V., Gawad, J.B., Kamble, M.D., Tauro, S.J., Bhandari, A.B., 2014. Promoieties used in prodrug design: a review. Ind. J. Pharm. Edu. Res. 48 (2), 35–40.

Praveen, B., Shrivastava, P., Shrivastava, S.K., 2009. *In-vitro* release and pharmacological study of synthesized valproic acid–dextran conjugate. Acta Pharm. Sci. 51, 169–176.

Rautio, J., Kumpulainen, H., Heimbach, T., Oliyai, R., Jarvinen, T., Savolainen, J., 2008. Prodrugs: design and clinical applications. Drug Discov. Nat. Rev. 7, 255–270.

Redasani, V.K., Bari, S.B., 2012. Synthesis and evaluation of mutual prodrugs of ibuprofen with menthol, thymol and eugenol. Eur. J. Med. Chem. 56, 134–138.

Rohini, N.A., Anupam, J., Alok, M., 2013. Polymeric prodrugs: recent achievements and general strategies. J. Antivirals Antiretrovirals S15, 1–12.

Vig, B.S., Huttunen, K.M., Laine, K., Rautio, J., 2013. Amino acids as promoieties in prodrug design and development. Adv. Drug Deliv. Rev. 65 (10), 1370–1385.

CHAPTER 5

Applications

5.1 APPLICATIONS OF PRODRUG DESIGNING

There are many barriers to drug design. the prodrug approach is used to overcome these barriers in order to improve the acceptability of available therapeutic agents. Overcoming these obstacles is helpful in the development of commercially available drug products. The designing of prodrugs has several applications. These applications are input to improve the problems associated with existing drugs and are responsible for raising the therapeutic effectiveness of drug molecules.

1. *Aesthetic properties*, such as odor, taste (in the case of pediatric use or when intended for oral administration), pain upon injection, gastrointestinal (GI) irritability of the new molecule.
2. *Drug formulation problems* namely, stability profile, undesirable physicochemical properties like solubility, polarity, partition coefficient, and pKa values, which preclude a drug's incorporation into a specific drug delivery system.

The utility of prodrugs to overcome the various *aesthetic and drug formulation problems* is discussed in the following sections.

5.1.1 Masking Taste and Odor

Taste and odor present problems, particularly with pediatric patients. One of the reasons for poor patient compliance, particularly in the case of children, is bitterness, acidity, or causticity of the drug. Masking of undesirable taste and odor is needed to overcome poor patient compliance. Two approaches can be utilized to overcome the bad taste of drugs. The first is reduction of drug solubility in saliva and the other is to lower the affinity of drug towards taste receptors. The undesirable taste arises due to adequate solubility and interaction of drug with taste receptors, which can be solved by lowering the solubility of drug or prodrug in saliva. Chloramphenicol, an extremely bitter drug, has been derivatized to chloramphenicol palmitate, a

Prodrug Design. DOI: http://dx.doi.org/10.1016/B978-0-12-803519-1.00005-2

sparingly soluble ester. It possesses low aqueous solubility, which makes it tasteless and later undergoes *in vivo* hydrolysis to active chloramphenicol by the action of pancreatic lipase.

Odor is another aesthetic concern for some drugs that are often volatile liquid or solids with significant vapor pressure that makes them difficult to formulate. A classic example is the volatile mercaptans used as tuberculostatic agents for the treatment of leprosy. Ethyl mercaptan has a boiling point of 25°C and a strong disagreeable odor. On the other hand, diethyl dithio isophthalate, a prodrug of ethyl mercaptan, has a higher boiling point and is relatively odorless.

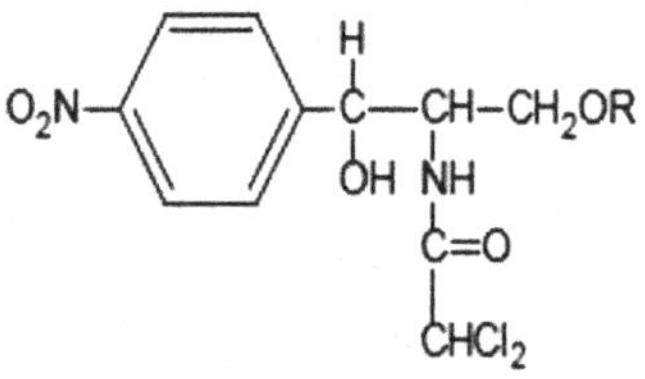

O_2N–C₆H₄–CH(OH)–CH(NH–C(=O)–$CHCl_2$)–CH_2OR

CH_3CH_2SH

R= H; Chloramphenicol

R= O=C-$(CH_2)_{14}$ CH_3; Chloramphenicol palmitate

Volatile with thiol odor

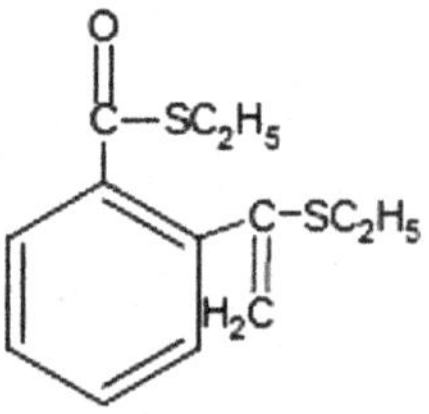

Relatively nontoxic with reduced thiol odor

5.1.2 Minimizing Pain at Injection Site

Pain at the injection site, especially after intramuscular injection, can be caused by drug precipitation and transfer of the drug into surrounding cells. It is caused by intramuscular injection mainly due to the weakly acidic nature or poor aqueous solubility of drugs. Such pain is usually associated with hemorrhage, edema, inflammation, and tissue necrosis. Some of these problems relate to dose vehicle composition or pH. For example, intramuscular injection of antibiotics like clindamycin and anticonvulsant drugs like phenytoin was found to bepainful due to poor aqueous solubility and could be overcome by making phosphate ester prodrugs respectively and maintaining the formulations at pH 12. The 50-fold greater solubility of the 2-phosphate ester prodrug of clindamycin results in little local pain or irritation on injection. The prodrug possesses little or no intrinsic antibacterial activity but is rapidly converted to the parent drug by the action of phosphatase in the body.

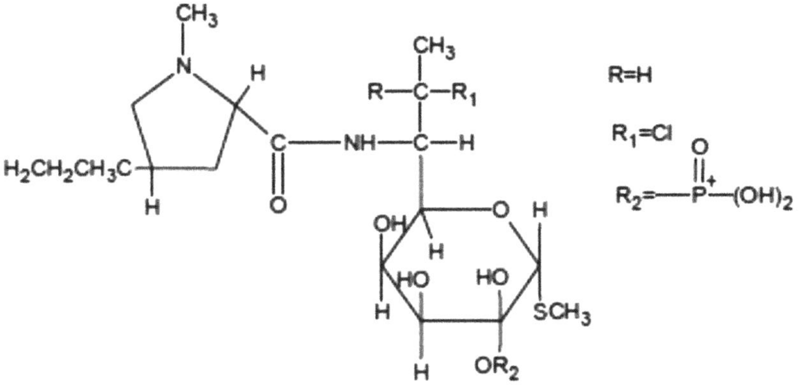

Clindamycin – 2 dihydrogen phosphate prodrug of clindamycin

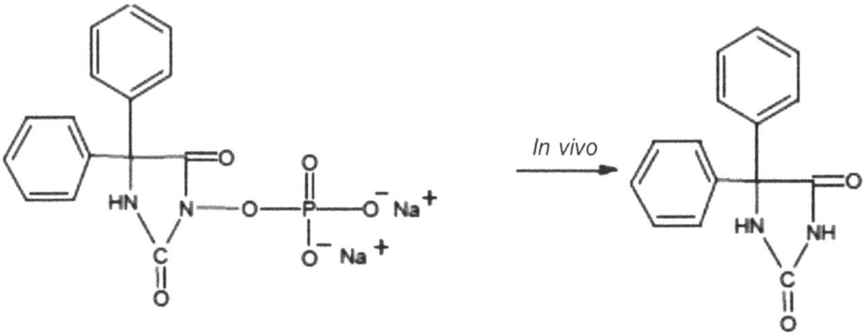

Phosphate ester of phenytoin

Phenytoin

Phenytoin and its prodrug

5.1.3 Alteration of Drug Solubility

Poor aqueous solubility can often prevent a drug from achieving its full therapeutic potential. For orally administered drugs, poor solubility in the intestinal contents would result in poor absorption. The prodrug approach can be used to increase or decrease the solubility of a drug, depending on its ultimate use. Many prodrugs designed to increase water solubility involve the addition of an ionizable promoiety to the parent molecule. Because charged molecules have greater difficulty crossing biological membranes, one must balance increased water solubility with the potential for decreased permeability. Prodrugs are also designed to improve aqueous solubility for parenteral administration. For example, chloramphenicol succinate and chloramphenicol palmitate, ester prodrugs of chloramphenicol, have enhanced and reduced aqueous solubility, respectively. On the basis of altered solubility, chloramphenicol sodium succinate prodrug is suitable for parenteral administration. The prodrug approach is also useful for better GI absorption. It was observed that sulindac, a prodrug of sulindac sulfide, being more water-soluble with sufficient lipophilicity, makes this drug suitable for oral administration.

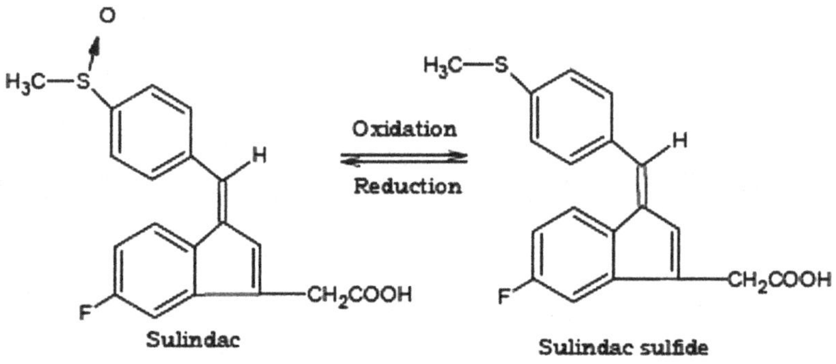

Prodrugs with altered solubility

5.1.4 Enhancement of Chemical Stability

Chemical stability is an extremely necessary parameter for every therapeutic agent to elicit its pharmacological activity for a longer duration. Many drugs are unstable and may either breakdown on prolonged storage or degrade rapidly on administration. Several drugs may decompose in the GI tract when used orally. A shelf life of at least 2 years is desirable, except for vaccines, cytotoxic agents, and other life-saving drugs. Although chemical instability can be solved to a greater extent by appropriate formulations, particularly enteric coatings, its failure necessitates the use of prodrug approach. The prodrug approach is based on the modification of the functional group responsible for the instability or by changing the physical properties of the drug, resulting in a reduction in the contact between the drug and the media in which it is unstable. This approach was successfully used to inhibit auto-aminolysis, which occurs due to the capability of the NH_2 group of side chain to attach to the lactam ring of other molecule, in an ampicillin molecule in concentrated solution it generates polymeric species of ampicillin. Making hetacillin, a prodrug of ampicillin formed by the reaction of acetone and ampicillin, ties up the amine group and thus inhibits auto-aminolysis.

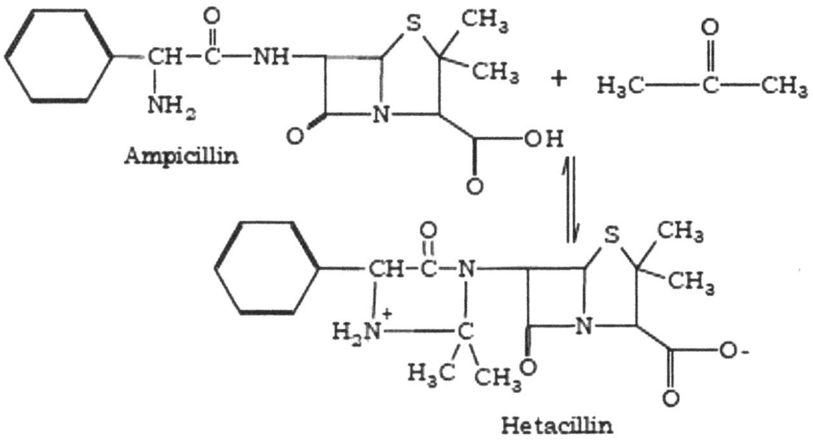

The prodrug approach can be successfully used to *overcome the various pharmacokinetic barriers*, thereby improving the therapeutic value of parent drug.

The principal barriers identified in the pharmacokinetic phase are:

1. Incomplete absorption of the drug from the delivery system or across biological barriers such as the GI mucosal cells and the blood–brain barrier.
2. Incomplete systemic delivery of an agent due to presystemic metabolism in the GI lumen mucosal cells and liver.
3. Toxicity problems associated with local irritation or distribution into tissue other than the desired target organ.
4. Poor site-specificity of the drug.

5.1.5 Prodrugs to Overcome Absorption Problems

Poor absorption of drug may be due to physicochemical properties of the drug itself. Bioavailability after oral dosing of various water-insoluble agents is often dissolution rate-limited, whereas the absorption of highly polar agents is often limited by their transport across the GI cell membrane. Since most drugs are absorbed by passive diffusion, a degree of lipophilicity is necessary for efficient absorption through the GI barrier. For highly polar compounds, the administration of less polar and more lipophilic prodrug promotes GI absorption. Also, many drugs are poorly absorbed into the central nervous system, eye, or through the skin due to their highly polar nature and a prodrug approach helps in overcoming these barriers.

5.1.5.1 Enhancement of Oral Absorption

Various therapeutic agents, such as water-soluble vitamins, structural analogu of natural purine and pyrimidine nucleoside, dopamine, antibiotics, such as ampicillin and carbenicillin, phenytoin, and cardiac glycoside, such as gitoxin, suffer from poor GI absorption. The prime cause of the poor absorption of these agents is their highly polar nature, poor lipophilicity, and/or metabolism during the absorption process. In contrast, gitoxin, a cardiac glycoside, has very poor oral bioavailability due to limited aqueous solubility. This problem could be manipulated successfully by using the prodrug approach. The absorption of water-soluble vitamin was enhanced by derivatization of thiolate ions to form lipid-soluble prodrugs. Dopamine was made useful by making its precursor L-dopa. Although L-dopa is highly polar, it is actively transported through a specific L-amino acid active transport mechanism and regenerates dopamine by decarboxylation.

This approach has been adopted with various penicillin antibiotics. Ampicillin is zwitterionic in the pH range of the GI tract and is only about 20–60% absorbed following oral dosing. Esterification of the carboxyl group of ampicillin to form the prodrugs pivampicillin, bacampicillin, or talampicillin alters the polarity of the molecule and successfully improves oral bioavailability.

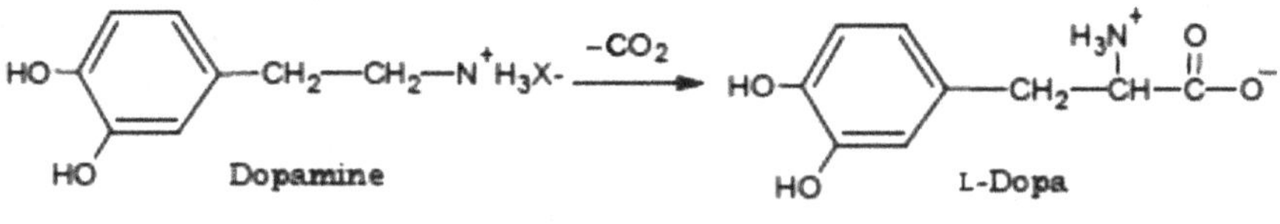

Dopamine $\xrightarrow{-CO_2}$ L-Dopa

Dopamine prodrugs

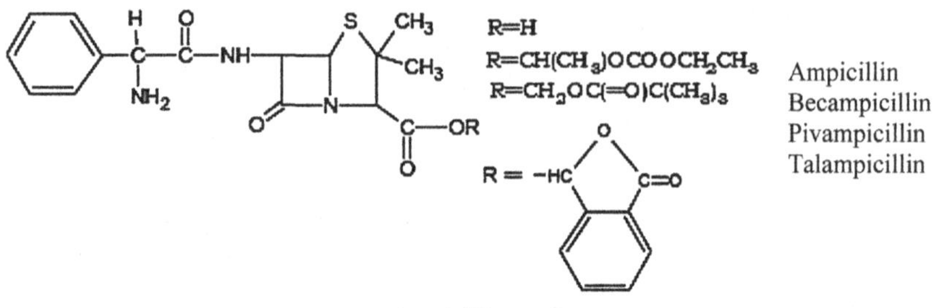

R=H — Ampicillin
R=$CH(CH_3)OCOOCH_2CH_3$ — Becampicillin
R=$CH_2OC(=O)C(CH_3)_3$ — Pivampicillin
R= –HC(phthalide) — Talampicillin

Ampicillin prodrugs

5.1.5.2 Enhancement of Ophthalmic Absorption

The usefulness of epinephrine as an adrenergic agent in the treatment of glaucoma is limited due to its highly polar nature. The dipivalyl derivative of epinephrine, formed by the acylation of phenolic hydroxyl groups, showed enhanced therapeutic effectiveness. Lipid solubility of dipivalyl derivatives is far superior to its parent compound, which facilitates its transport through a lipoidal barrier during corneal absorption.

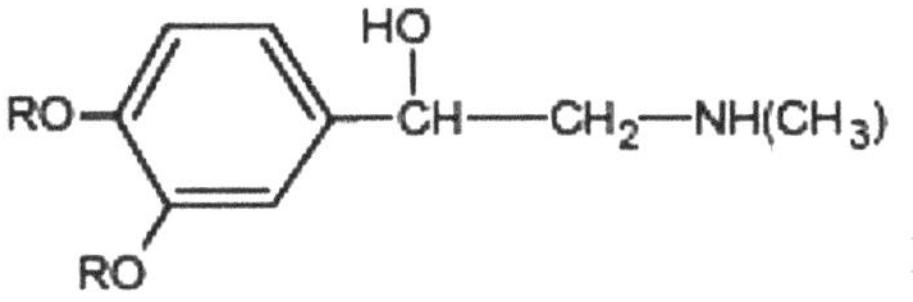

R = H — Epinephrine

R = $COCH_3$ — Dipivalyl derivative of epinephrine

Prodrugs used in enhancement of ophthalmic absorption

5.1.5.3 Enhancement of Percutaneous Absorption

Mefenide and corticosteroid are used in the treatment of inflammatory, burn therapy, allergic, and pruritic conditions, but have limited application due to poor percutaneous absorption. It was observed that mefenide hydrochloride and mefenide acetate salt showed better responses than the parent drug. However, due to the much stronger basic nature of acetate ion, the equilibrium is forced towards the formation of parent drug rapidly in comparison to hydrochloride salt. The problem of poor percutaneous absorption of corticosteroid was overcome by making various ester prodrugs.

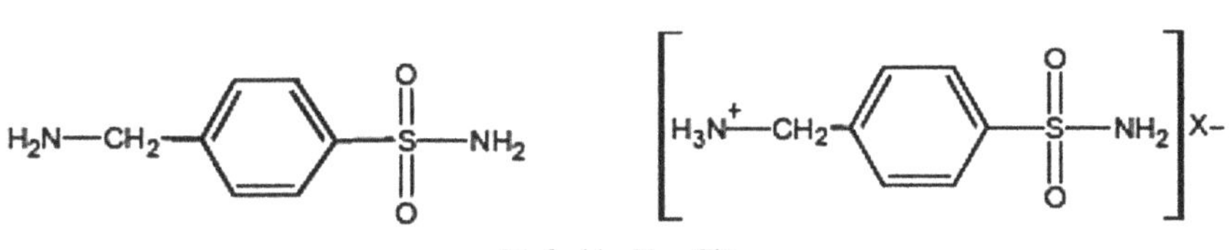

Mefenide; X = Cl^-

Mefenide acid salt; X = CH_3COO^-

5.1.6 Prevention of Presystemic Metabolism

Many drugs are efficiently absorbed from the GI tract but undergo presystemic first-pass metabolism or chemical inactivation before reaching the systemic circulation. Phenolic moiety, oxidative N- and O-dealkylation, ester cleavage, and peptide degradation are responsible for the presystemic metabolism of various drugs. In fact, two types of drug fall into this category. The first are drugs rapidly degraded by the acid condition of the stomach and the drugs of the second category degrade due to enzymes present in the GI mucosa and liver. Enzymatic degradation is perhaps of greater significance than chemical degradation. Following oral administration, a drug must pass through two metabolizing organs, i.e., liver and GI mucosa, before reaching the general circulation as shown in following figure.

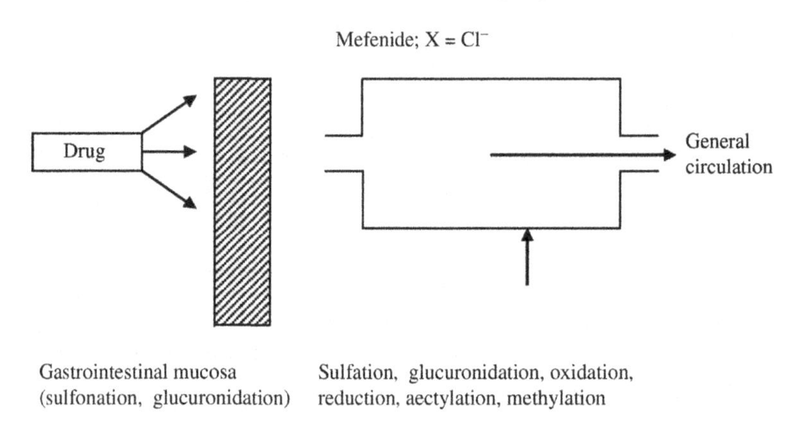

Rapid metabolism of drugs in these organs is termed the first-pass effect. The first-pass metabolism of a drug can be prevented if the functional group susceptible to metabolism is protected temporarily by derivatization. Alternatively, manipulation of the drug to alter its physicochemical properties may also alter the drug–enzyme complex formation.

The prodrug approach was successfully used to overcome the problem of considerable metabolism of steroidal drugs, propranolol, dopamine, morphine, and catecholamines, by making acetylated derivates of various steroids, 17α, 21-acetonides of various corticosteroids, hemisuccinate ester of propranolol, L-dopa prodrug in the case of dopamine, diacetyl prodrug of morphine and ibuterol and bitoterol prodrugs of terbutaline and N-(t butyl arternol), respectively. All the prodrugs, besides enhancing absorption and bioavailability, protect the therapeutic agent from metabolism.

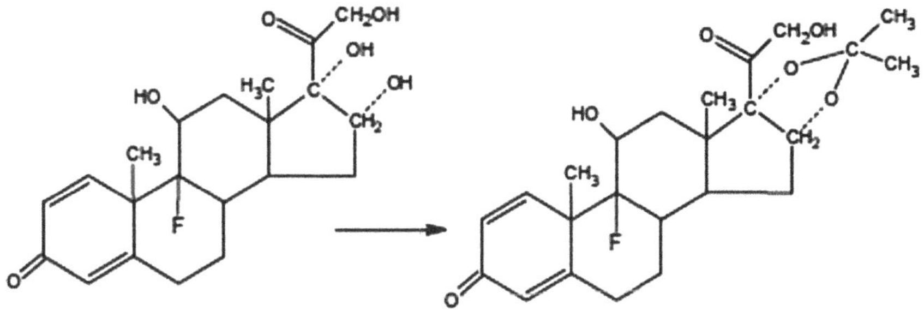

Triamcinolone

16 α, 17α acetonides of triamcinolone

5.1.7 Longer Duration of Action

Drugs with a short half-life require frequent dosing with conventional dosage forms to maintain adequate plasma concentration of the particular drug. Frequent dosing for drugs rapidly cleared from the body results in a sharp peak and valley effect. In plasma level time profile, and consequently patient compliance, is often poor. The peak and valley effect can be minimized by drug delivery at a controlled and predictable rate, such as zero order delivery. Prolongation of duration of action of a drug can be accomplished by the prodrug approach and can take two forms. First, the input of drug in to the body can be controlled by a prodrug/drug delivery formulation complex, which by design releases drug at a controlled rate at the absorption site, followed by conversion to drug prior to or just after absorption. Second, a prodrug can be designed wherein the conversion to the parent drug becomes the release rate-limiting factor in the systemic milieu.

The present approach is most useful in the case of neuroleptic drugs to avoid large fluctuations in plasma levels. This could be successfully achieved by administering hepatanoate and decanoate esters of fluphenazine in sterile sesame oil. Similarly, the problem of testosterone is overcome by administering 17-propionate ester, 17-phenylacetylate, and 17-cypionate ester of testosterone in oil vehicle.

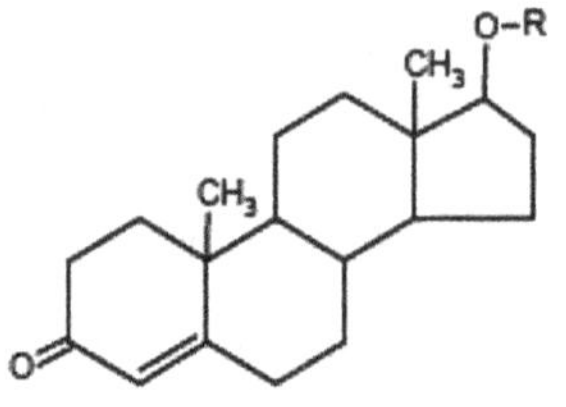

R = H; Testosteron

R = $OC-CH_2-CH_2-$(cyclopentyl)

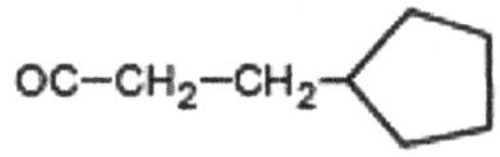

Cypionate ester

Testosteron prodrugs

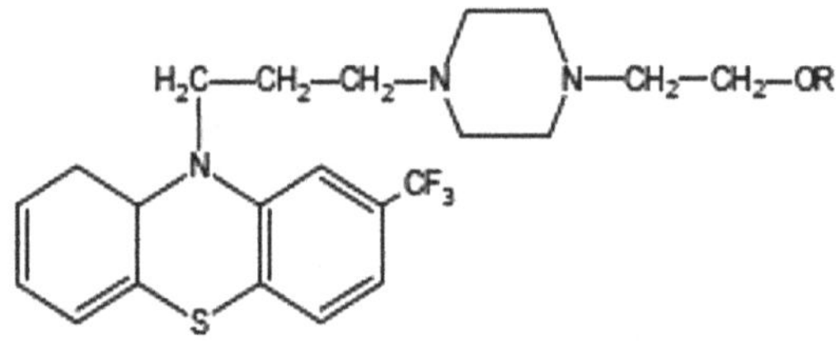

R= H; Fluphenazine

R = COC_6H_{13} ; Fluphenazine heptanoate

R = COC_9H_{19} ; Fluphenazine decanoate

Fluphenazine prodrugs

5.1.8 To Diminish Local and Systemic Toxicity of Drugs/ Reduction of GI Irritability

One of the desired properties in drug design and targeting is to have therapeutic activity without toxicity. It seems very difficult unless site-specific delivery of drug is achieved. Some examples exist where a prodrug approach has resulted in a reduction in toxicity. Various nonsteroidal anti-inflammatory drugs, such as salicylic acid and indometacin, severely damage the GI mucosa due to the presence of a free carboxylic group. Aspirin has already been mentioned as a less corrosive prodrug of acetylsalicylic acid. Several mechanisms of action have been proposed to explain these, including injury to gastric mucosa through direct contact of drug, damage due to stimulation of gastric acid secretion by circulating drug, and interference with the protective GI mucosal layer. The mucosal layer is relatively resistant to proteolytic enzyme activity and protects the underlying epithelial tissue.

Similarly, the nonsteroidal anti-inflammatory sulindac shows reduced local GI irritation compared with corresponding sulfide. The anti-inflammatory activity and GI intolerance of many anti-inflammatory drugs relate to their inhibition of prostaglandin synthesis. As a sulfoxide, sulindac shows little or no such activity and this may account for the reduced local GI irritation compared to active sulfide. Sulindac shows little or no such pharmacological activity but it is rapidly reduced to sulfide after absorption.

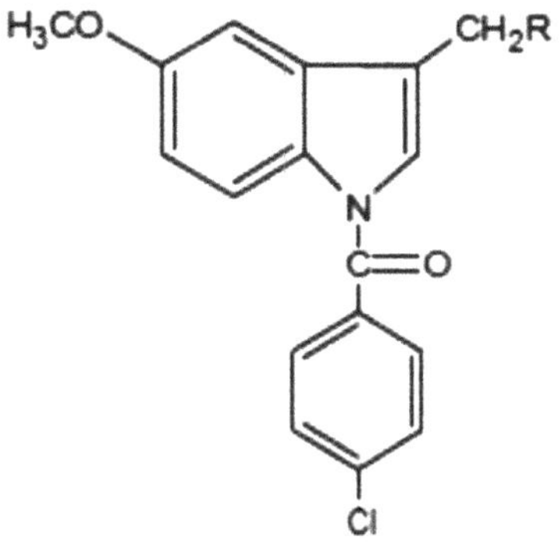

R= H; Indomethacin / R = CHO; Aldehyde prodrug of indomethacin

5.1.9 Site-Specific Drug Delivery

Site-specific drug delivery attempts to obtain very precise and directs effect at the site of action, without subjecting the rest of the body to a significant level of active agent. Certain limitations to this approach have been highlighted as follows.

1. Adequate accessibility of prodrug to the target site
2. Distribution, relative activity, and specificity of reconverting enzymes
3. Parent drug leakage from the target site
4. Access of parent drug to the target site within the target organ
5. Pharmacokinetics of the prodrug and drug molecules
6. Toxicity of prodrug moiety and its metabolites
7. Lack of information on the selective location of potential reconverting enzymes.

L-3,4-dihydroxyphenylalanine (L-Dopa) has been used in the treatment of Parkinson's, a neurogenerative disorder resulting from the degeneration of dopaminergic neurons in the basal ganglia. Treatment with neurotransmitter dopamine is impracticable due to lack of absorption and metabolism in the intestine and blood–brain barrier and is concentrated in the dopaminergic and adrenergic nerve terminals, where it is converted to dopamine, noradrenaline, or adrenaline.

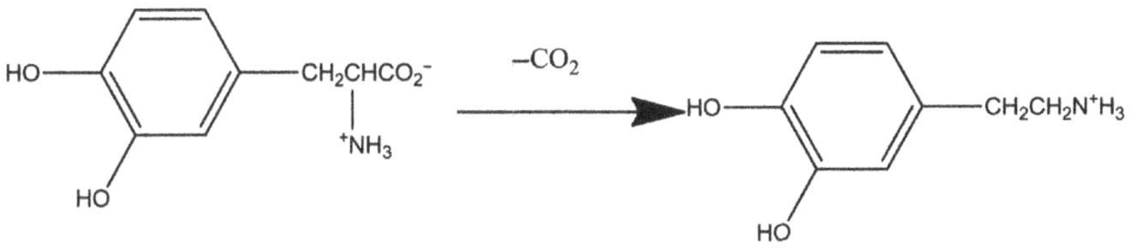

5.1.10 Prodrug for Slow and Prolonged Release (Sustained Drug Action)

A common strategy in the design of slow-release prodrug is to make long-chain aliphatic esters, because these esters hydrolyze slowly, and to inject them intramuscularly. Fluphenazine has a shorter duration of action (6–8 h), but prodrug fluphenazine decanoate has a duration of activity of about a month.

REFERENCES

Delgado, J.N., Remers, W.A., 1998. Drug latentiation and Prodrugs. Wilson and Gisvold's Text Book of Organic, Medicinal and Pharmaceutical Chemistry, tenth ed. Lippincott Williams and Wilkins Philadelphia, pp. 123–138.

Hansch, C., Sammes, P.G., Taylor, J.B., 2005. Comprehensive Medicinal Chemistry: "Principles of Pharmacokinetics and Metabolism", vol. 5. Pergamon Press, Oxford England, pp. 122–133.

Rautio, J., Kumpulainen, H., Heimbach, T., Oliyai, R., Jarvinen, T., Savolainen, J., 2008. Prodrugs: design and clinical applications. Drug Discov. Nat. Rev. 7, 255–270.

Souzan, Y., Dhiren, R.T., 2007. Prodrugs: absorption, distribution, metabolism, excretion (ADME) issues. Prodrugs Biotechnol. Pharm. Aspects 5, 1043–1081.

Stella, V.J., Nti Addae, K.W., 2007. Prodrug strategies to overcome poor water solubility. Adv. Drug Deliv. Rev. 30 (59), 677–694.

Stella, V.J., Charman, W.N., Naringrekar, V.H., 1985. Prodrugs. Do they have advantages in clinical practice? Drugs 29, 455–473.

CHAPTER 6

Role in Drug Discovery

Before using a prodrug approach, the following criteria should be met: (i) there is a real problem that is worth addressing; (ii) the prodrug is transient; (iii) the components of the prodrug do not cause extra toxicity; and (iv) the prodrug should be cheap and simple to make. Several common promoieties that can be used to make prodrugs of various functional groups were reviewed, including acyl and "soft" alkyl groups. It was noted that chemical stability of prodrugs may be different to enzymatic stability, for example, the chemical stability of aliphatic carboxylic esters in buffered aqueous solutions increases with increasing chain length of the aliphatic acid. Enzymatic hydrolysis of esters, however, increases initially with chain length and decreases when the chain length is longer than six to seven carbons.

Prodrug design can be highly effective for solving many of the stability, solubility, permeability, and targeting problems in drug discovery and development. A number of successful examples of prodrug design were presented that resulted in pharmaceutical products, either already commercialized or in late-stage development. In addition to improving oral bioavailability, prodrugs are increasingly used for targeting purposes, including site-specific activation and delivery of anticancer drugs to tumor tissues through transporters, tumor- or tissue-specific enzymes, and gene therapy.

6.1 ROLE OF CYTOCHROME P450 (CYP450)

Cytochrome P450 (CYP450) plays an important role in the metabolism and detoxification of exogenous substances, including drugs taken orally. Knowledge of the CYP450 system is critical in understanding drug metabolism and drug interactions, and early identification of CYP450 substrates, inhibitors, and inducers, as well as possible reactive metabolites, has become an integral part of the lead optimization process. The biochemical and genetic aspects of CYP450 enzymes help in CYP450-mediated prodrug activation processes (Hu, 2004).

Prodrug Design. DOI: http://dx.doi.org/10.1016/B978-0-12-803519-1.00006-4

6.2 CASE STUDIES IN PRODRUG DESIGN

6.2.1 Ximelagatran as a Double Prodrug of Melagatran to Increase Permeability

Melagatran was identified in the 1990s as a potent direct inhibitor of thrombin and platelet aggregation. However, its oral bioavailability is only approximately 5% and is greatly reduced when dosed with food. This is due to a chemical part present in the structure of the compound. Due to the presence of two strongly basic groups and one carboxylic acid group, it exists as a zwitterion at intestinal pH. Troy C Sarich (AstraZeneca Plc, USA) presented the discovery and development of Ximelagatran, which was developed as a double prodrug of melagatran in order to increase permeability while maintaining good pharmacological properties. Thus, the carboxylic acid group was converted to an ester and the imidine moiety was hydroxylated to reduce its basicity. Ximelagatran is not charged at intestinal pH and is 170-fold more lipophilic and 80-fold more permeable than its parent drug (Gudmundsson, 2007).

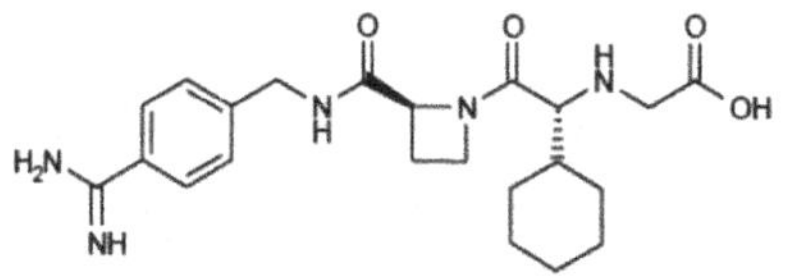

Melagatran

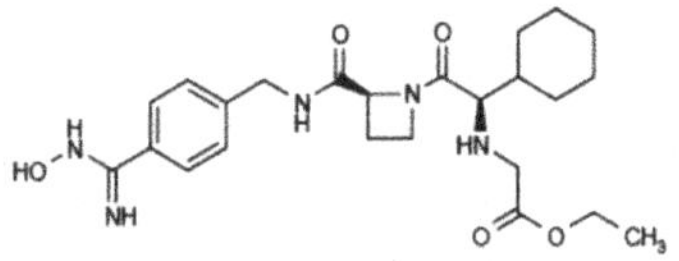

Ximelagatran

6.2.2 Fosamprenavir: A Soluble Prodrug Leads to a Better Product

Amprenavir is an HIV protease inhibitor used to treat HIV infections in adults and children. This drug was approved by the FDA in 1999. The limited water-solubility necessitates the use of a soft gel formulation and multiple pills for a single dose. Dan Todd (GlaxoSmithKline Plc, USA) converse Fosamprenavir as a prodrug of amprenavir with improved aqueous solubility. After screening several potential prodrugs *in vitro* and *in vivo*, the phosphate prodrug fosamprenavir was selected because of its high water-solubility, solution and solid-state stability and its rapid conversion to the parent drug on the apical side of epithelium prior to absorption of the parent drug. The prodrug Fosamprenavir was approved by the FDA in October 2003 for use in combination with other antiretroviral agents for the treatment of HIV infection in adults.

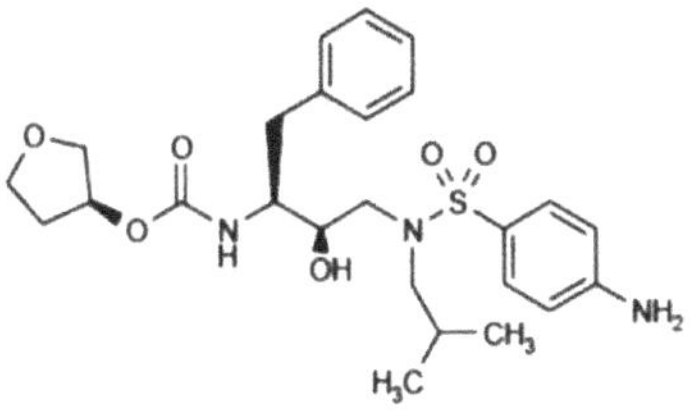

Amprenavir

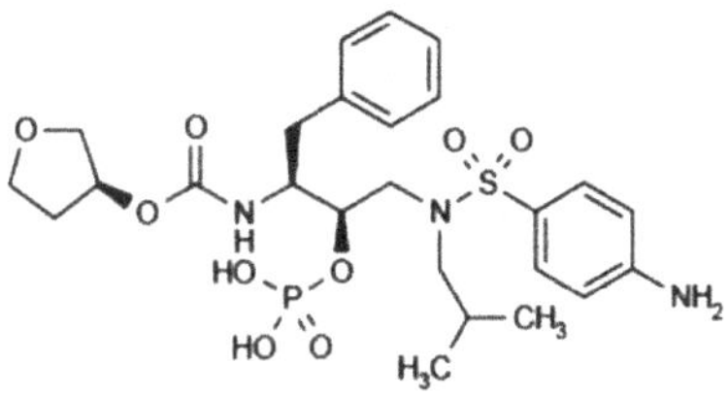

Fosamprenavir

6.2.3 Ampiroxicam: An Anti-Inflammatory Prodrug Agent

Piroxicam is a nonsteroidal anti-inflammatory drug that can cause serious gastrointestinal bleeding, perforation, and ulceration. Anthony Marfat (Pfizer Inc., USA) discussed the use of the prodrug concept that led to the discovery of ampiroxicam as an anti-inflammatory agent that reduces the local or topical component of gastric irritation. This has been done by masking the enolic hydroxyl group of piroxicam. Over 225 derivatives of piroxicam were made and screened for cyclooxygenase inhibitory activity, solution bioavailability in rats, and solid-state stability and absorption in dogs. Ampiroxicam was found to meet all decision criteria and was developed further.

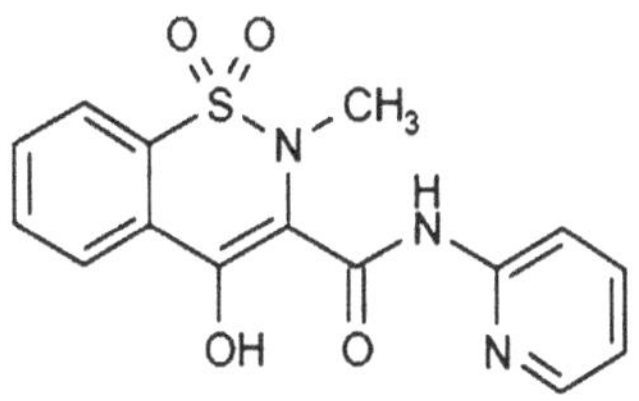

Piroxicam

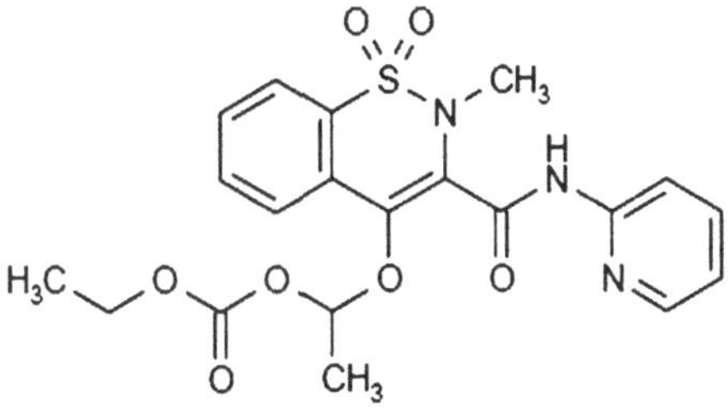

Ampiroxicam

6.2.4 Tenofovir Disoproxil Fumarate: A Successful Anti-HIV Agent

Tenofovir is an acyclic nucleoside phosphonate that undergoes phosphorylation to form tenofovir diphosphate; the latter is a potent and selective inhibitor of viral reverse transcriptase that effectively blocks viral replication. However, tenofovir is a dianion at physiological pH with a low partition coefficient, and its oral bioavailability is low and erratic in animal studies. Reza Oliyai (Gilead Sciences Inc., USA) reported the design and development of tenofovir disoproxil fumarate, a prodrug designed to improve the permeability of tenofovir across biological membranes. Several ester prodrugs of tenofovir were evaluated.

Tenofovir disoproxil, the *bis*-isopropoxyl carbonate derivative, was the most chemically stable in solution and quickly converted back to tenofovir in the presence of tissue homogenates. The oral bioavailability of the prodrug for tenofovir is found to be approximately 30%.

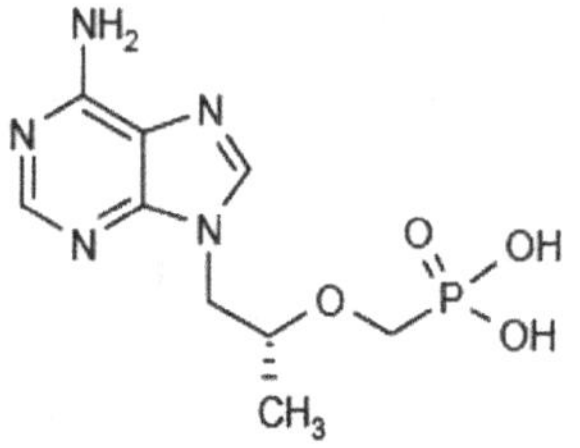

Tenofovir

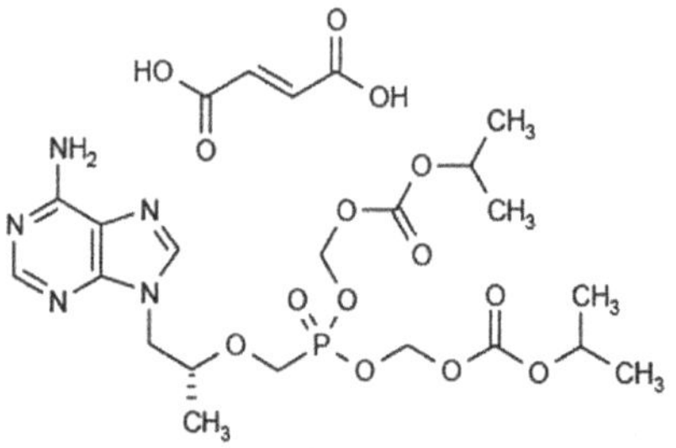

Tenofovir disoproxil fumarate

6.2.5 Olmesartan Medoxomil: A Prodrug of Olmesartan

Olmesartan is a durable, specific, and competitive nonpeptide angiotensin II receptor antagonist used to treat hypertension. To achieve better oral bioavailability its ester prodrug, olmesartan medoxomil, was developed. Olmesartan medoxomil is suitable for once-daily oral administration and is available in tablet form (Bianca, 2007).

6.2.6 Valacyclovir: A Prodrug of Acyclovir

Valacyclovir is an L-valyl ester prodrug of acyclovir that is used for the treatment of herpes, varicella zoster, and cytomegaloviruses. Valacyclovir was developed to increase the oral absorption and plasma levels of acyclovir. Increased plasma concentrations of acyclovir are important in maintaining antiviral activity, especially in immune-compromised patients and in the treatment of less sensitive viruses. Suboptimal exposures can lead to more resistant viral strains. To achieve high enough exposures, acyclovir must be dosed intravenously or in multiple high doses (Melissa and Olafur, 2007).

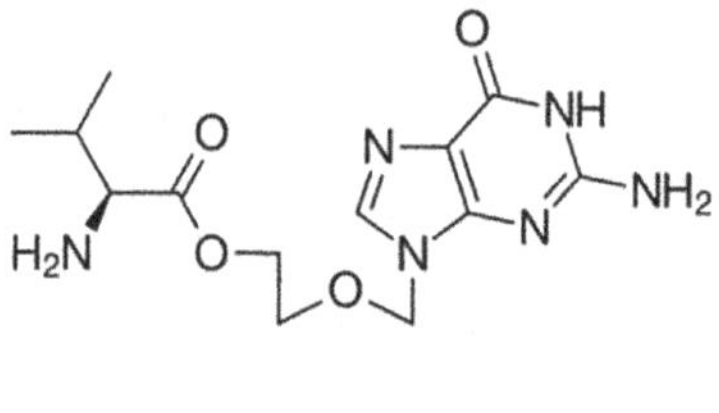

Valacyclovir

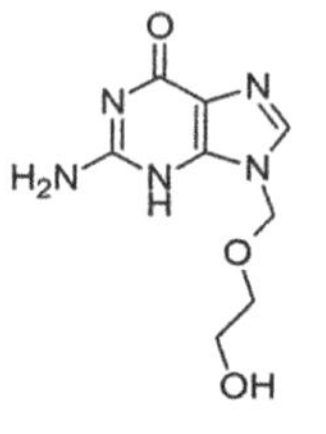

Acyclovir

6.2.7 ZYN001: A Prodrug that Enables Effective Transdermal Delivery

Most cannabinoid therapies have drawbacks and limitations in safety and efficacy due to the fact that they are administered orally, including low bioavailability, inconsistent plasma levels, and significant first-pass liver metabolism. First-pass liver metabolism refers to the process by which the liver breaks down therapeutics ingested directly or indirectly through the gastrointestinal system, such as through oral or oral mucosal delivery methods, allowing only a small amount of drug to emerge into the circulatory system. In our preclinical animal studies, ZYN001 demonstrated effective skin permeation with sustained delivery and rapid conversion of ZYN001. These preclinical studies suggest increased bioavailability, consistent plasma levels, and the avoidance of first-pass liver metabolism. Alternatively, transdermal therapeutics are absorbed through the skin directly into the systemic circulation, avoiding first-pass liver metabolism and potentially enabling lower dosage levels of active pharmaceutical ingredients and rapid and reliable absorption with high bioavailability.

REFERENCES

Bianca, M.L., 2007. Case study: olmesartan medoxomil: a prodrug of olmesartan. Prodrugs Biotechnol. Pharm. Aspects 5, 1305–1312.

Gudmundsson, O.S., 2007. Case study: ximelagatran: a double prodrug of melagatran. Prodrugs Biotechnol. Pharm. Aspects 5, 1395–1402.

Hu, L., 2004. Prodrugs: effective solutions for solubility, permeability and targeting challenges. IDrugs 7 (8), 736–742.

Melissa, D.A., Olafur, S.G., 2007. Case study: valacyclovir: a prodrug of acyclovir. Prodrugs Biotechnol. Pharm. Aspects 5, 1369–1376.

CHAPTER 7

Work Reported

As stated previously, prodrugs are an inactive form of drug which need conversion in the body to one or more of their active metabolites. The metabolite form is the one which is active and is capable of producing the desired reaction. It offers advantages over the active form in being more stable, having better bioavailability, and other desirable pharmacokinetic properties, or less side effects and toxicity.

7.1 SOME COMMONLY USED PRODRUGS

The following section deals with the examples of commonly synthesized prodrugs along with their therapeutic applications and mode of biotransformation. The final destination in prodrug design is its conversion into parent active drug.

1. Carisoprodol is metabolized into meprobamate. Until 2012, carisoprodol had not been a controlled substance, but meprobamate was classified as a potentially addictive controlled substance that can produce dangerous and painful withdrawal symptoms upon discontinuation of the drug.
2. Enalapril is bioactivated by esterase to the active enalaprilat.
3. Valaciclovir is bioactivated by esterase to the active aciclovir.
4. Fosamprenavir is hydrolyzed to the active amprenavir.
5. Levodopa is bioactivated by DOPA decarboxylase to the active dopamine.
6. Chloramphenicol succinate ester is used as an intravenous prodrug of chloramphenicol, because pure chloramphenicol does not dissolve in water.
7. Psilocybin is dephosphorylated to the active psilocin.
8. Heroin is deacetylated by esterase to the active morphine.
9. Molsidomine is metabolized into the active compound nitric oxide.
10. Paliperidone is an atypical antipsychotic for schizophrenia. It is the active metabolite of risperidone.

Prodrug Design. DOI: http://dx.doi.org/10.1016/B978-0-12-803519-1.00007-6

11. Prednisone, a synthetic corticosteroid drug, is bioactivated by the liver into the active drug prednisolone, which is also a steroid.
12. Primidone is metabolized by cytochrome P450 enzymes into phenobarbital, which is the major, and phenylethylmalonamide, which is the minor metabolite.
13. Dipivefrine, given topically as an antiglaucoma drug, is bioactivated to epinephrine.
14. Lisdexamfetamine is metabolized in the small intestine to produce dextroamphetamine at a controlled (slow) rate for the treatment of attention deficit hyperactivity disorder.
15. Diethylpropion is a diet pill that does not become active as a monoamine releaser or reuptake inhibitor until it has been N-dealkylated to ethylpropion.
16. Fesoterodine is an antimuscarinic that is bioactivated to 5-hydroxymethyl tolterodine, the principal active metabolite of tolterodine.
17. Tenofovir disoproxil fumarate is an anti-HIV drug (NtRTI class) that is bioactivated to tenofovir (PMPA).

7.2 LIST OF PRODRUGS AND THEIR RESPECTIVE ACTIVE FORMS

Prodrugs are used in certain conditions when the drug in action cannot reach the desired site and the inactive form helps the active form in reaching the desired site of action. The table below is a list of prodrugs along with their active forms that help in conversion and assessing the desired activity.

Sr. No.	Prodrug	Active Form
1	Levodopa	Dopamine
2	Enalapril	Enalaprilat
3	S-methyldopa	α Methyl norepinephrine
4	Dipivefrine	Epinephrine
5	Sulindac	Sulfide metabolites
6	Hydrazide (mono amino oxidase inhibitor)	Hydrazine derivatives
7	Proguanil	Proguanil triazine
8	Prednisone	Prednisolone
9	Becampicillin	Ampicillin
10	Sulfasalazine	5-Aminosalicyclic acid
11	Cyclophosphamide	Aldophosphamide, phosphormide mustard
12	5-Fluorouracil	Fluororidine monophosphate
13	Mercaptopurine	Methylmercaptopurine ribonucleotide

7.3 LIST OF COMMERCIALLY AVAILABLE PRODRUGS

The following is a list of some commercially available prodrugs. These are the approved prodrugs that are marketed by pharmaceutical companies. These are representative examples as it is not possible to produce the entire list here.

1. Oseltamivir and zanamivir are currently licensed worldwide for influenza treatment and chemoprophylaxis. Both drugs require twice-daily administration for 5 days for treatment. A new influenza drug, laninamivir, and its prodrug form, laninamivir octanoate or laninamivir prodrug, which are long-acting neuraminidase inhibitors, are introduced.
2. Hepsera (adefovir dipivoxil) and Viread (tenofovir disoproxil fumarate) are the antivirals from Gilead Sciences used in the treatment of hepatitis and AIDS, respectively.
3. Valcyte is the prodrug of Valganciclovir, an antiviral agent marketed by Roche Holding AG.
4. Benicar is the prodrug of Olmesartan, an antihypertensive agent from Sankyo Co. Ltd.
5. Dynastst is the prodrug of Paracoxib an analgesic agent from Pharmacia (Pfizer).
6. Aquavan (GPI-15715) is a water-soluble prodrug of Propofol from Guilford/ProQuest Limited.
7. Ximelagatran is prodrug of Exanta and BIBR-1048 is a prodrug of Dabigatran etexilate, an oral anticoagulant drugs developed by AstraZeneca.

The Occurrence of Prodrugs Amongst the World's Top Selling Pharmaceuticals		
Prodrug (Trade name) and Therapeutic Area	**Functional Group**	**Prodrug Strategy**
Proton Pump Inhibitors		
Esomeprazole (Nexium)	Formation of active sulfonamide form	Bioprecursor prodrugs that are converted into their respective active sulfonamide forms site-selectively in the acidic conditions of the stomach
Lansoprazole (Prevacid)		
Pantoprazole (Protonix)		
Rabeprazole (Aciphex)		
Antiplatelet Agent		
Clopidogrel (Plavix)	Formation of active thiol	Bioprecursor prodrug that selectively inhibits platelet aggregation

Antiviral Agent		
Valacyclovir (Valtrex)	L-Valyl ester of acyclovir	Bioconversion by valacyclovir hydrolase. Transported predominantly by hPEPT1. Oral bioavailability improved from 12–20% (acyclovir) to 54% (valacyclovir)
Hypercholesterolemia		
Fenofibrate (Tricor)	Isopropyl ester of fenofibric acid	Lipophilic ester of fenofibric acid
Antiviral Agent		
Tenofovir disoproxil (Atripla)	*Bis*-(isopropyl-carbonyl oxymethyl) ester of tenofovir	Bioconversion by esterases and phosphodiesterases
Psychostimulant		
Lisdexamfetamine (Vyvanse)	L-Lysyl amide of dextroamphetamine	Bioconversion by intestinal or hepatic hydrolases. Reduced potential for abuse due to prolonged release of active drug
Influenza		
Oseltamivir (Tamiflu)	Ethyl ester of oseltamivir carboxylate	Improved bioavailability compared with oseltamivir carboxylate, allowing oral administration
Hypertension		
Olmesartan medoxomil (Benicar)	Cyclic carbonate ester of olmesartan	Improved bioavailability compared with olmesartan, allowing oral administration
Immunosuppressant		
Mycophenolate mofetil (Cell Cept)	Morpholinyl ethyl ester of mycophenolic acid	Improved oral bioavailability with less variability
Glaucoma		
Latanoprost (Xalatan)	Isopropyl ester of latanoprost acid	Bioconversion by esterases. Improved lipophilicity to achieve better ocular absorption and safety

CONCLUSION

There are several ways to improve the safety and efficacy of existing drugs. A prodrug is intended to overcome barriers to utility, such as solubility, bioavailability, or lack of site-specificity, through a chemical approach rather than through formulation.

An important financial consideration in drug development is the patent protection and market exclusivity that the drug may receive. One difficulty in bringing a prodrug to market is that the determination of market exclusivity for a prodrug has been evolving in the Food and Drug Administration (FDA) over the years. This process has included the interpretation and re-interpretation of regulations establishing what constitutes a prodrug for the purpose of defining a period of market exclusivity. If a prodrug is determined to be a new chemical entity, it is eligible for 5 years of market exclusivity.

Market exclusivity is a vital factor in determining the marketability and profit potential of any drug. By revisiting its previous decisions, the FDA has ensured consistency in exclusivity determinations. Understanding recent exclusivity decisions surrounding approved prodrugs are a key to gaining insight into probable future decisions regarding prodrug exclusivity. Professional counsel in this area is prudent for any drug development program.

Academic Press is an imprint of Elsevier
125, London Wall, EC2Y 5AS
525 B Street, Suite 1800, San Diego, CA 92101-4495, USA
225 Wyman Street, Waltham, MA 02451, USA
The Boulevard, Langford Lane, Kidlington, Oxford OX5 1GB, UK

Notices
Knowledge and best practice in this field are constantly changing. As new research and experience broaden our understanding, changes in research methods or professional practices, may become necessary.

Practitioners and researchers must always rely on their own experience and knowledge in evaluating and using any information or methods described herein. In using such information or methods they should be mindful of their own safety and the safety of others, including parties for whom they have a professional responsibility.

To the fullest extent of the law, neither the Publisher nor the authors, contributors, or editors, assume any liability for any injury and/or damage to persons or property as a matter of products liability, negligence or otherwise, or from any use or operation of any methods, products, instructions, or ideas contained in the material herein.

ISBN: 978-0-12-803519-1

British Library Cataloguing-in-Publication Data
A catalogue record for this book is available from the British Library

Library of Congress Cataloging-in-Publication Data
A catalog record for this book is available from the Library of Congress

For Information on all Academic Press publications
visit our website at http://store.elsevier.com/

Printed in the United States
By Bookmasters